NEED to KNOW

HIGHER HUMAN BIOLOGY

Key facts at your fingertips

Graham Moffat

Billy Dickson

HODDER
GIBSON
AN HACHETTE UK COMPANY

Although every effort has been made to ensure that website addresses are correct at time of going to press, Hodder Gibson cannot be held responsible for the content of any website mentioned in this book. It is sometimes possible to find a relocated web page by typing in the address of the home page for a website in the URL window of your browser.

Hachette UK's policy is to use papers that are natural, renewable and recyclable products and made from wood grown in well-managed forests and other controlled sources. The logging and manufacturing processes are expected to conform to the environmental regulations of the country of origin.

Orders: please contact Bookpoint Ltd, 130 Park Drive, Milton Park, Abingdon, Oxon OX14 4SE. Telephone: (44) 01235 827827. Fax: (44) 01235 400401. Email education@bookpoint.co.uk. Lines are open from 9 a.m. to 5 p.m., Monday to Saturday, with a 24-hour message answering service. You can also order through our website: www.hoddereducation.co.uk. Hodder Gibson can also be contacted directly at hoddergibson@hodder.co.uk.

ISBN: 978 1 5104 5116 2

© Graham Moffat, Billy Dickson 2019

First published in 2019 by
Hodder Gibson, an imprint of Hodder Education,
An Hachette UK Company
211 Vincent Street
Glasgow G2 5QY

www.hoddereducation.co.uk

Impression number 10 9 8 7 6 5 4 3 2 1

Year 2023 2022 2021 2020 2019

Typeset by Aptara, Inc., India
Printed in Spain

A catalogue record for this title is available from the British Library.

MIX
Paper from
responsible sources
FSC™ C104740

Contents

Getting the most from this book

This *Need to Know* guide is designed to help you throughout your Higher Human Biology as a course companion to your learning but also as a revision aid to be used in preparation for course assessments and for your final examination.

You need to know

Each Key Area begins with a list of the learning outcomes adapted from the SQA course specification for Higher Human Biology. These summarise the general themes of learning within the topic.

The Key Area continues with bullet notes covering the success criteria for each learning outcome. These contain the emboldened vocabulary and phrasing needed to ensure exam success. Some terms are highlighted in green if they are defined as selected key terms.

Key terms

The **selected key terms** are only a small sample of the terms you need to know for your exam. There are many more terms you need to know, which you'll find **emboldened** and defined throughout bullet points in the book.

Exam tips

Exam tips focus on areas that are tricky and are often asked about in the exam. These may contain a hint, a memory aid or a note of what to watch for.

Synoptic links

These are references to other Key Areas in the book to show where related knowledge for Higher Human Biology can be found – these are *always* worth checking out.

Techniques

These are experimental techniques with which you are expected to be familiar for your exam – we give a brief outline of each technique and its purpose.

Do you know?

Questions at the end of each Key Area can be used to test yourself on the main knowledge needed. These are in the form of extended-response questions worth between 3 and 8 marks each. Give yourself about 2 minutes for each mark, so a 5-mark question should take you about 10 minutes.

Mark your own work here: hoddereducation.co.uk/ needtoknow/answers

Area assessment

A group of ten structured questions, worth 60 marks, designed to give you examination practice across a whole area of the human biology course.

Mark your own work here: hoddereducation.co.uk/ needtoknow/answers

1 Human cells

1.1 Division and differentiation in human cells

Somatic and germline cells

- A **somatic cell** is any cell in the body other than cells involved in reproduction.
- Somatic cells are **diploid**, which means that they contain two sets of chromosomes that exist in **homologous** pairs. Each member of a homologous pair carries the same sequence of genes.
- Diploid cells in humans have 23 pairs of **homologous chromosomes**.
- Somatic **stem cells** divide by **mitosis** to form more somatic cells, as shown in Figure 1.1.
- **Germline cells** are cells involved in reproduction. Germline cells are the stem cells that divide to form gametes (sperm and ova).
- Germline stem cells divide by mitosis and by **meiosis**.
  - □ Their nuclei divide by mitosis to produce more germline cells and maintain the diploid number of chromosomes in daughter cells, as shown in Figure 1.1.
  - □ In meiosis, the cell undergoes two divisions, firstly separating homologous pairs of chromosomes and secondly separating pairs of **chromatids** to produce **haploid** gametes, as shown in Figure 1.2.
- The haploid gametes contain 23 single chromosomes.
- When gametes fuse at fertilisation, the zygote formed is diploid and has 23 pairs of homologous chromosomes, as shown in Figure 1.3.

Key terms

Somatic cell Diploid body cell.

Homologous chromosomes Sexually reproducing organisms have cells in which chromosomes occur as homologous pairs – each parent contributed one member of each pair and the members of the pairs carry the same sequence of genes.

Mitosis Cell division which maintains the diploid chromosome number and is used in growth and repair.

Germline cell Reproductive cell that divides and develops to form gametes.

Meiosis Cell division used in haploid gamete production.

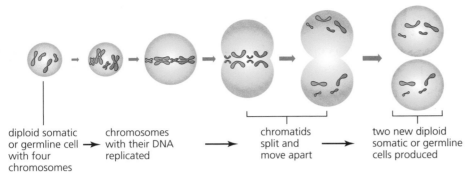

diploid somatic
or germline cell → chromosomes
with four
chromosomes

chromosomes
with their DNA
replicated

chromatids
split and
move apart

two new diploid
somatic or germline
cells produced

Figure 1.1 Stages of mitosis in a somatic cell. Note that only four chromosomes are shown but that in humans the diploid number is 46

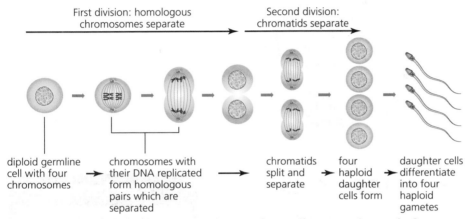

First division: homologous
chromosomes separate

Second division:
chromatids separate

diploid germline
cell with four
chromosomes

chromosomes with
their DNA replicated
form homologous
pairs which are
separated

chromatids
split and
separate

four
haploid
daughter
cells form

daughter cells
differentiate
into four
haploid
gametes

Figure 1.2 Stages of meiosis in male germline cells. Note that only four chromosomes are shown but that in humans the diploid number is 46

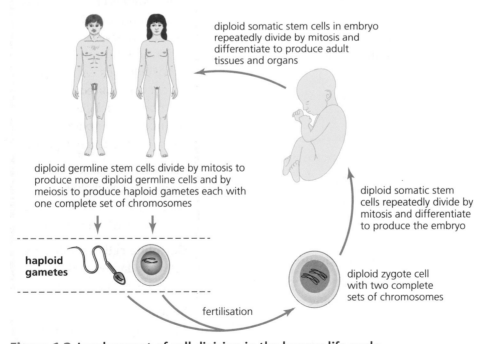

diploid somatic stem cells in embryo
repeatedly divide by mitosis and
differentiate to produce adult
tissues and organs

diploid germline stem cells divide by mitosis to
produce more diploid germline cells and by
meiosis to produce haploid gametes each with
one complete set of chromosomes

diploid somatic stem
cells repeatedly divide by
mitosis and differentiate
to produce the embryo

**haploid
gametes**

fertilisation

diploid zygote cell
with two complete
sets of chromosomes

Figure 1.3 Involvement of cell division in the human life cycle

Cellular differentiation

- Cellular differentiation is the process by which a cell switches on certain genes to express proteins characteristic of that type of cell.
- Other genes in the cell are switched off and so are not expressed, as shown in Figure 1.4.

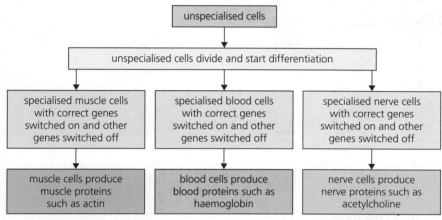

Figure 1.4 **Differentiation of unspecialised cells into muscle, red blood cells and nerve cells**

- Differentiated cells can carry out specialised functions because of the proteins they contain.
- For example, red blood cells have a large surface area for exchange of oxygen and have haemoglobin to carry oxygen; neurons have long fibres which carry electrical messages and can release neurotransmitters.

Stem cells

- Stem cells are unspecialised somatic cells that can divide by mitosis to make copies of themselves (self-renew) and to make cells that differentiate into specialised cells of one or more types.
- **Embryonic stem cells** and **tissue stem cells** are the two types of stem cell in humans.
  - □ Embryonic stem cells in the very early embryo are pluripotent, which means they could differentiate into all the cell types that make up the individual. This is because all the genes in embryonic stem cells have the potential to be switched on, as shown in Figure 1.5(a).
  - □ Tissue stem cells are multipotent, which means they can differentiate into all of the types of cell found in their particular tissue type. They are involved in the growth, repair and renewal of the cells found in that tissue type, as shown in Figure 1.5(b). For example, blood stem cells located in bone marrow can become red blood cells, platelets, phagocytes and lymphocytes.

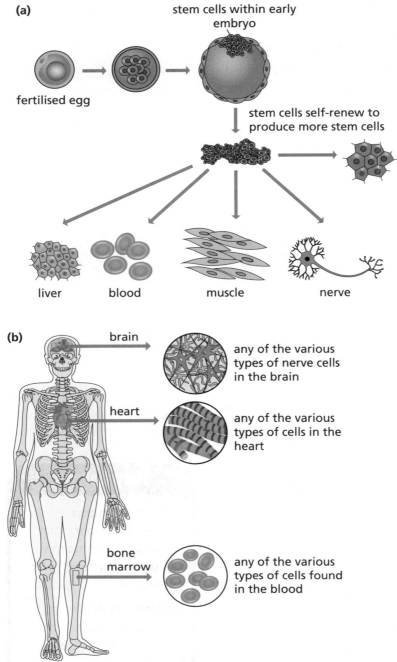

Figure 1.5 (a) Embryonic and **(b)** tissue stems cells

Uses of stem cells

- Stem cells have both **therapeutic** and **research** uses.
  - □ Therapeutic uses involve treatments for patients including the repair of damaged or diseased organs or tissues; for example, in the repair of corneal tissue damaged by cataracts, or in the repair and regeneration of skin that has been burned.

Synoptic links

You can read more about phagocytes in Key Area 3.5 (page 85) and lymphocytes in Key Area 3.6 (page 87).

Key term

Therapeutic Used as part of a medical treatment to repair damaged or diseased organs or tissues.

☐ In research, stem cells can be used as model cells to study how diseases develop. These model cells can also be used to test the effects of new drugs on cells in general.

☐ Stem cell research can also provide information on how cell processes such as cell growth, differentiation and gene regulation work.

Ethical issues with stem cells

■ Embryonic stem cells can be cultured in the laboratory. They self-renew in the culture to provide a supply of these cells for medical use.

■ Embryonic stem cells can provide effective treatments for disease and injury. However, this raises ethical issues because obtaining them involves the destruction of an early embryo.

Exam tip

Ethical issues are tricky – remember that an ethical issue is about the morality of a decision or an action. Stem cell treatments raise ethical issues because of the use of embryonic stem cell therapies. Is it right to destroy an embryo to obtain stem cells even although doing so may save the life of a patient?

Cancer cells

■ Cancer cells are abnormal cells that divide excessively because they do not respond to regulatory signals. This excessive division results in a mass of abnormal cells called a **tumour**.

■ Cells within the tumour may fail to attach to each other, spreading through the body where they may form secondary tumours, as shown in Figure 1.6.

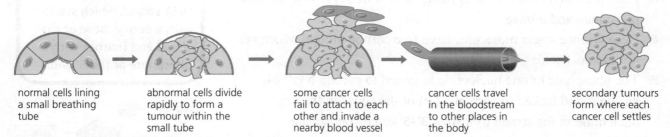

| normal cells lining a small breathing tube | abnormal cells divide rapidly to form a tumour within the small tube | some cancer cells fail to attach to each other and invade a nearby blood vessel | cancer cells travel in the bloodstream to other places in the body | secondary tumours form where each cancer cell settles |

Figure 1.6 Stages in the formation of a tumour and the production of secondary tumours

Do you know?

1 Give an account of embryonic and tissue stem cells. [5]

2 Give an account of somatic and germline cells. [5]

3 Give an account of stem cell research and the therapeutic use of stem cells. [4]

1.2 Structure and replication of DNA

You need to know

- the features that make up the structure of a DNA molecule
- how DNA replicates, including the roles of DNA polymerase, ligase and primers
- the purpose of the different heat treatments in the polymerase chain reaction (PCR)
- the practical applications of PCR

Structure of a DNA molecule

- DNA is a long, double-stranded molecule wound into the shape of a **double helix**.
- Each strand of the double helix is made up of repeating chemical units called nucleotides.
- Each individual nucleotide is made up of a **deoxyribose** sugar, a phosphate and a **base**.
- Deoxyribose sugar molecules have five carbon atoms, which are numbered 1 to 5.
- The phosphate of one nucleotide is joined to carbon 5 (5′) of its sugar and linked to carbon 3 (3′) of the sugar in the next nucleotide in the strand to form a **3′–5′ sugar–phosphate backbone**, as shown in Figure 1.7.
- Each strand of the double helix has a sugar–phosphate backbone with a 3′ end that starts with a deoxyribose sugar molecule and a 5′ end that finishes with a phosphate.
- A nucleotide has one of four different bases called adenine (A), guanine (G), thymine (T) and cytosine (C).
- The nucleotides of one strand of DNA are linked to the nucleotides on the second strand through their bases – the bases form pairs joining the strands.
- The bases pair in a complementary way: adenine always pairs with thymine, and guanine always pairs with cytosine.

Key terms

Nucleotide Component of DNA consisting of a deoxyribose sugar, a phosphate group and a base.

3′–5′ The direction of a DNA strand, which starts with a deoxyribose at the 3′ end and finishes with a phosphate at the 5′ end.

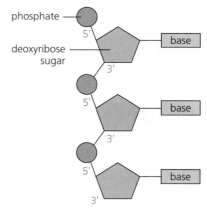

Figure 1.7 Short section of one strand of DNA showing how the sugars and phosphates link to form a 3′–5′ sugar–phosphate backbone

- Base pairs are held together by weak hydrogen bonds, as shown in Figure 1.8(a).
- The two strands of a DNA molecule run in opposite directions and are said to be **antiparallel** to each other, as shown in Figure 1.8(b).

Key terms

Complementary Applies to the specific base-pairing in DNA in which adenine always pairs with thymine, and guanine always pairs with cytosine.

Antiparallel Parallel strands in DNA which run in opposite directions. Each strand has a sugar–phosphate backbone with a 3′ end that starts with a deoxyribose molecule and a 5′ end that finishes with a phosphate.

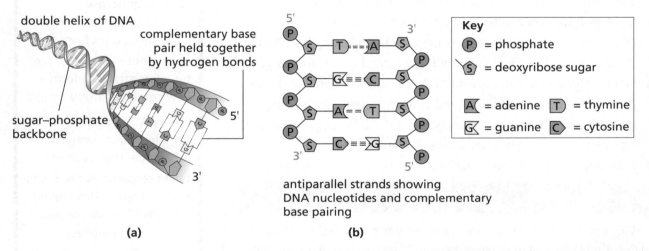

Figure 1.8 Features of a DNA molecule: (a) double-helix structure with strands linked by hydrogen bonds between complementary base pairs; (b) details of antiparallel strands and complementary base-pairing

Exam tip

DNA strands are a bit like lanes of traffic on a road – they are essentially the same but run in opposite directions. In your exam you may be asked what 'antiparallel' means. Remember the phrase, 'same but run in opposite directions'. One runs from the 3′ to the 5′ end and the other runs in the opposite direction.

Function of DNA

- The DNA of an organism encodes its heritable genetic information as a genetic code.
- The base sequence along one strand of a DNA molecule forms the genetic code.

Synoptic links

You can read more about the importance of the genetic code in Key Areas 1.3 and 1.5 (pages 15 and 22).

The process of DNA replication

- Replication is the process by which DNA molecules can direct the synthesis of identical copies of themselves.
- A copy is needed for each of the daughter cells produced in cell division.
- Prior to cell division, DNA is replicated by the enzyme **DNA polymerase**.
- DNA is unwound and hydrogen bonds between complementary bases are broken to form a fork with two template strands.
- One template is the **leading strand** and the other is the **lagging strand**.
- **Primers** bind to the template strands. A primer is a short complementary strand of nucleotides which binds to the 3' end of the leading template DNA strand. On the lagging strand, primers are bound at intervals on the strand as 3' sites are exposed.
- DNA polymerase needs these primers to start replication. DNA polymerase adds complementary DNA nucleotides to the deoxyribose (3') end of the new DNA strand that is forming.
- DNA polymerase can only add DNA nucleotides in one direction, from its 3' end towards its 5' end. This results in the leading strand being replicated continuously and the lagging strand being replicated in fragments, as shown in Figure 1.9.
- Fragments of DNA are joined together by the enzyme **ligase**.

Exam tip

In your exam you could be asked why DNA must replicate. DNA replication occurs prior to cell division because the daughter cells produced each need all the genetic information the parent cell had so that they can carry out all of their functions and code for all of the proteins required.

Key terms

DNA polymerase Enzyme that adds free DNA nucleotides from the deoxyribose (3') end of the template strand to form a new complementary strand.

Primer A short complementary strand of nucleotides that binds to a target sequence at the 3' end of the template DNA strand, allowing DNA polymerase to add DNA nucleotides.

Ligase Enzyme that joins DNA fragments during replication of the lagging strand.

Exam tip

Many students find it tricky to explain why the antiparallel nature of DNA affects how DNA polymerase synthesises a copy of its template strands. DNA polymerase can only replicate DNA in a 3' to 5' direction and only the leading strand runs in that direction; the polymerase has to wait until a fragment of the lagging strand is unzipped to allow replication of the fragment.

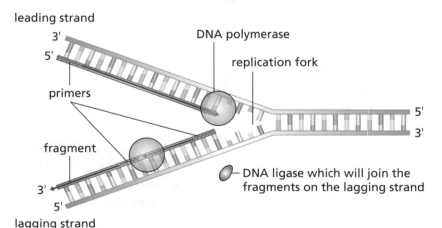

leading strand

3'
5'
primers
fragment
3'
5'
lagging strand

DNA polymerase
replication fork
5'
3'

DNA ligase which will join the fragments on the lagging strand

Figure 1.9 DNA polymerase replicating both strands of a DNA molecule and DNA ligase about to join fragments on the lagging strand

The polymerase chain reaction

- The **polymerase chain reaction (PCR)** is a laboratory technique for the amplification of target sequences of DNA using complementary primers for specific target sequences of interest.
- PCR primers are short strands of nucleotides that are complementary to specific target sequences at the two 3′ ends of the region of DNA to be amplified.
- A DNA molecule is heated to 92–98°C to break all the hydrogen bonds and completely separate the two strands. It is then cooled to between 50°C and 65°C to allow primers to bind to target sequences. Finally, heating to 70–80°C allows a heat-tolerant DNA polymerase to replicate the target DNA sequence by adding complementary DNA nucleotides to each strand.
- Repeated thermal cycles of heating and cooling amplify the target sequence of DNA, as shown in Figure 1.10.

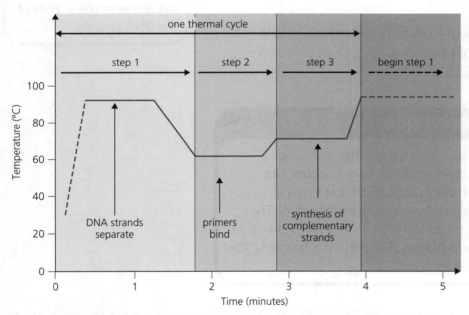

Figure 1.10 The temperatures and events in one thermal cycle of the PCR process

- Each thermal cycle of PCR doubles the number of copies of the DNA present, as shown in Figure 1.11.

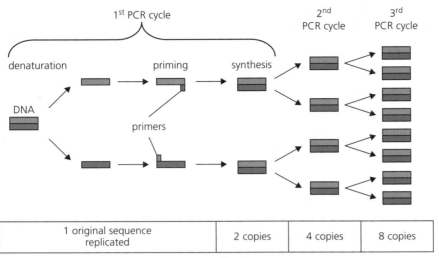

Figure 1.11 Amplification of DNA by a doubling of the number of molecules at each cycle of the PCR process

Applications of PCR

- PCR can amplify DNA for a variety of applications. The amplified DNA can be used to produce DNA profiles unique to each individual.
- PCR can be used in forensics to identify individuals to help solve crimes, in the settlement of paternity disputes and to identify genetic disorders in individuals.

Technique

Gel electrophoresis is used to make DNA profiles from DNA sequences that have usually been amplified from samples. The amplified DNA is treated with enzymes that cut it in specific places to produce a set of fragments unique to an individual. The fragments are stained and then separated in an agarose gel using an electric current to give patterns of lines. The pattern is unique to the individual whose DNA is being analysed.

Do you know?

1 Give an account of the structure of a single DNA nucleotide. [3]
2 Give an account of the structure of a DNA molecule. [5]
3 Describe the main steps in DNA replication. [5]
4 Describe the main steps in the polymerase chain reaction (PCR). [4]
5 Give *two* examples of the applications of PCR. [2]

1.3 Gene expression

You need to know

■ that gene expression involves the transcription and translation of DNA sequences
■ the role of RNA polymerase in the transcription of DNA into primary mRNA transcripts
■ how tRNA is involved in the translation of mRNA into a polypeptide at a ribosome
■ that different proteins can be expressed from one gene, as a result of alternative RNA splicing
■ that amino acids are linked by peptide bonds to form polypeptides

Genes

■ Gene expression is the process by which specific genes are activated to produce a required protein.
■ Genes are expressed to produce proteins which determine the **phenotype** of an individual. As well as gene expression, environmental factors such as diet also influence phenotype.
■ Only a fraction of the genes in any cell are expressed at any time in that cell's life.
■ Gene expression involves the **transcription** and **translation** of specific DNA sequences, as shown in Figure 1.12.

Key terms

Phenotype Features of an organism including its appearance and behaviour.

Messenger RNA (mRNA) Carries a complementary copy of the genetic code from the nucleus to a ribosome.

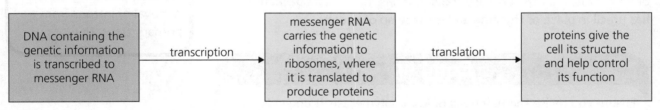

Figure 1.12 Summary of gene expression

RNA

■ Transcription and translation involve three types of RNA called mRNA, tRNA and rRNA.
■ RNA is single-stranded and is composed of nucleotides containing ribose sugar, phosphate and a base.
■ The RNA bases are cytosine, guanine, adenine and uracil (which replaces thymine), as shown in Figure 1.13.
■ Uracil in RNA is complementary to adenine.

Key terms

Transfer RNA (tRNA) Carries a specific amino acid from the cytoplasm to a ribosome.

Ribosomal RNA (rRNA) Combines with protein to form the structure of a ribosome.

- Transcription occurs in the nucleus of a human cell.
- Each triplet of bases on the DNA within a gene is called a **codon**.
- Most codons code for specific amino acids, but there are also **start** and **stop codons** at the beginning and end of the gene sequence. A start codon indicates where transcription should begin and a stop codon signals where it should end.
- In transcription, the enzyme **RNA polymerase** moves along DNA, unwinding the double helix and breaking the hydrogen bonds between the bases.
- RNA polymerase then synthesises a **primary transcript** of mRNA from free RNA nucleotides by aligning them against their complementary base-pair partners. Uracil is aligned against adenine, as shown in Figure 1.14.
- Like DNA, the primary transcript has groups of three bases called codons.

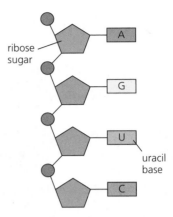

Figure 1.13 Short strand of RNA and its main features

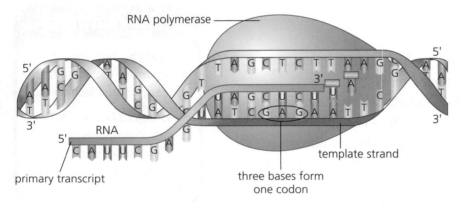

Figure 1.14 Transcription of DNA to produce a primary transcript of RNA. Note that the primary transcript has ribose in its backbone, has uracil in place of thymine and is made up of codons

Key terms

Codon Sequence of three bases on mRNA which specifies an amino acid.

RNA polymerase Enzyme involved in synthesis of primary transcripts from DNA.

Primary transcript Molecule made when DNA is transcribed.

Introns Non-coding sequences of DNA or RNA.

RNA splicing Joining of exons following the removal of introns from a primary transcript.

Exam tip

Ensure you revise the functions of RNA polymerase. It unwinds and breaks hydrogen bonds between DNA bases and adds complementary nucleotides to the primary transcript – it is really several enzymes in one.

Alternative splicing

- The **introns** of the primary transcript are non-coding regions of the RNA and are identified and removed from the transcript by enzymes in a process called **RNA splicing** (see Figure 1.15).

- The **exons** are coding regions of the RNA. The order of the exons is unchanged during splicing and the spliced exons form a **mature mRNA transcript**.
- Different mature mRNA transcripts are produced from the same primary transcript depending on which exons are retained, as shown in Figure 1.15.
- Different proteins can be expressed from one gene as a result of alternative RNA splicing.
- The mature mRNA transcripts each carry a complementary copy of the DNA sequences from the nucleus to the cytoplasm, to be translated into proteins by ribosomes.

Exam tip

In your exam you may be asked to explain the significance of alternative RNA splicing. Remember that different mature transcripts (mature mRNA) can be produced so that different proteins can be produced from a single gene. One gene; many proteins.

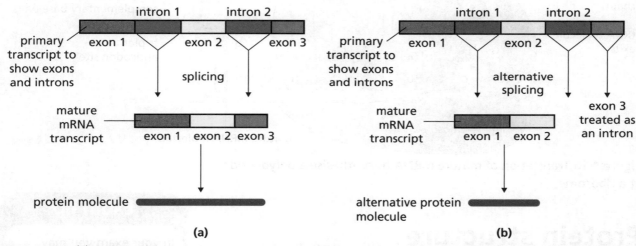

Figure 1.15 (a) Splicing of a primary transcript to produce a mature mRNA – note that exons are retained in the same order that they occur in the primary transcript; (b) an alternative splicing of the same primary transcript – again note that the order of exons is unchanged but an alternative protein is produced

Translation

- **Ribosomal RNA (rRNA)** molecules are folded together with proteins to form a ribosome.
- Each mature mRNA transcript joins with a ribosome, as shown in Figure 1.16.
- Translation occurs at the ribosome and involves the synthesis of a polypeptide by **transfer RNA (tRNA)** molecules.
- tRNA is folded into shape by complementary base pairs which hold the molecule in a particular shape.

Exam tip

Make sure that you can describe the structure of tRNA, including details about the sugar ribose and presence of uracil in nucleotides, as well as talking about anticodons, amino acid attachment sites and folding due to complementary base-pairing.

- A tRNA molecule has an exposed triplet of bases at one end called an **anticodon** and an **attachment site** for the specific amino acid encoded by its anticodon at the other end.
- Each different tRNA molecule carries a specific amino acid to the ribosome.
- Translation begins at the start codon of the mRNA and ends at the stop codon.
- tRNA anticodons align and bond to mRNA codons by complementary base-pairing, bringing their amino acids into a sequence and so translating the genetic code.
- **Peptide bonds** then form to join the amino acids together to form a **polypeptide**.
- Following polypeptide formation, tRNA exits the ribosome to collect further amino acids.

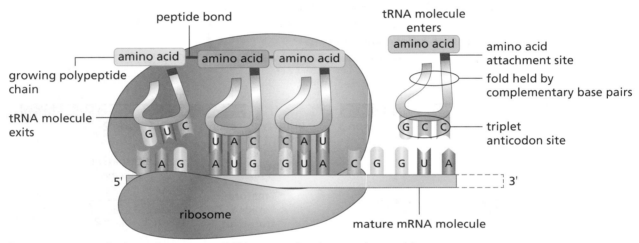

Figure 1.16 Translation of mature mRNA to synthesise a polypeptide at a ribosome

Protein structure

- Amino acids are linked by **peptide bonds** to form polypeptides.
- The polypeptide chains fold to form the three-dimensional shape of a protein, held together by hydrogen bonds and other interactions between individual amino acids, as shown in Figure 1.17.

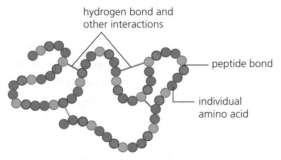

Figure 1.17 Various bonds hold amino acids in the three-dimensional shape of a protein molecule

Exam tip

In your exam you may be asked to explain how DNA carries the genetic code and the importance of codons and anticodons in translation. Remember that it is a triplet code, where each triplet of bases codes for a specific amino acid. The tRNA anticodons are complementary to the mRNA codons and this ensures that the amino acids are joined in the correct sequence to synthesise the protein.

- Proteins can form a large variety of shapes. Shape determines protein function.
- Some proteins act as enzymes. Some proteins make up the structure of cells, such as those involved in membranes.

Exam tips

- In your exam you may be asked to explain the importance of the three-dimensional shape of a protein molecule. Remember that protein function depends on its shape. Shape is important in the active site of enzymes, receptor proteins and hormones.
- There are several enzymes mentioned throughout the course – DNA polymerase, RNA polymerase, ligase and ATP synthase. Be sure you know what each of these does.

Synoptic link

You can read more about the importance of protein shape in Key Area 1.6 (page 23), where the shape of active sites is related to the function of enzymes.

Do you know?

1 Give an account of the transcription of DNA.   [4]
2 Describe the events leading to the formation of a mature mRNA transcript following transcription.   [4]
3 Give an account of the translation of mature RNA.   [4]
4 Describe the structure of proteins.   [3]

1.4 Mutations

You need to know

- that mutations are changes to DNA that can result in no protein or an altered protein being synthesised
- that single-gene mutations involve the alteration of a DNA nucleotide sequence as a result of the substitution, insertion or deletion of nucleotides
- that chromosome structure mutations include duplication, deletion, inversion and translocation

Mutations

- Mutations are spontaneous, random and rare changes in DNA that can result in either failure to synthesise a specific protein or the synthesis of an altered protein.

Gene mutations

- **Single-gene mutations** involve changes in DNA nucleotide sequences which can result from the **substitution, insertion** or **deletion** of nucleotides in the gene, as shown in Figure 1.18.
- Substitution (Figure 1.18(a)) does not affect the position of codons but could result in a wrong amino acid being used in the protein produced or, if it resulted in a stop codon, the protein produced could be shorter. Both insertion and deletion are **frame-shift mutations** and disrupt the sequence of codons from the mutation point downstream in the gene.
- Nucleotide insertions or deletions (Figure 1.18 (b) and (c)) result in frame-shift mutations because the code is read in triplets, so all of the codons after the mutation are affected. The consequence is that all of the amino acids after the mutation will be changed in the protein expressed. This has a major effect on the structure of the protein produced.

> ### Key term
>
> **Frame-shift mutation** A nucleotide insertion or deletion that causes all of the codons and all of the amino acids after the mutation to be changed.

> ### Exam tip
>
> Remember that single-gene mutations all involve changes in nucleotide sequence. The effects of a deletion or insertion on the protein produced can be tricky to describe. A good word to include in any answer is 'frame-shift', and remember to say that the effects are major because they change all the amino acids after the mutation.

	(a) substitution	(b) insertion	(c) deletion
original sequence	T G G C A G	T G G C A G	T G G C ⚡ G
sequence following mutation	T G G T A G	T G G T C A G	T G G C G
	one nucleotide replaced by a different one	an extra nucleotide put into the sequence	a nucleotide removed and lost from the sequence

Figure 1.18 The effect of different gene mutations on the sequence of nucleotides in a gene

- Nucleotide substitutions include missense, nonsense and splice-site mutations.
  - □ **Missense mutations** affect amino acid codons and result in one amino acid being changed for another. This may result in a non-functional protein or have little effect on the protein.
  - □ **Nonsense mutations** result in an amino acid codon being changed to give a premature stop codon, which results in a shorter protein and will have a major effect on the protein's function.
  - □ **Splice-site mutations** affect the nucleotide sequence at the splice site. This may result in an intron being retained and/or an exon not being included in the mature transcript. Any protein produced will be seriously affected by this.

> ### Key terms
>
> **Missense mutation** Single-nucleotide substitution mutation that results in one amino acid being changed for another.
>
> **Nonsense mutation** Single-gene mutation that results in a premature stop codon being produced, which results in a shorter protein.

Chromosome mutations

■ **Chromosome structure mutations** that involve alterations to the structure of a chromosome include **deletion, inversion, translocation** and **duplication**, as shown in Figure 1.19.
  □ Deletion is where a section of a chromosome is removed and lost from the genome (gene D in Figure 1.19(a)).
  □ Inversion is where a section of chromosome has its order in the chromosome reversed (genes B, C and D in Figure 1.19(b)).
  □ Translocation is where a section of a chromosome is added to a chromosome, not from its homologous partner. In Figure 1.19(c), genes at A and B on one chromosome and those at M, N and O on a non-homologous chromosome are swapped.
  □ Duplication is where a section of a chromosome is repeated by being added from its homologous chromosome partner. In Figure 1.19(d) genes at B and C from one homologous chromosome are passed to its homologous partner.
■ These substantial changes in chromosome mutations often make them lethal.

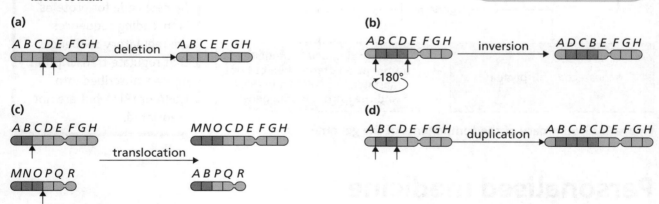

Figure 1.19 The effects of different chromosome mutations on the sequence of genes in the genome

1.5 Human genomics

You need to know

- that the human genome is made up of genes and other DNA sequences that do not code for proteins
- that in genomic sequencing, the sequence of nucleotide bases can be determined for individual genes and entire genomes
- about pharmacogenetics and personalised medicine

Structure of the human genome

- The **genome** of an organism is the entire hereditary information encoded in DNA.
- The human genome is made up of genes, which are DNA nucleotide sequences that code for proteins, and other DNA sequences that do not code for proteins.
- **Coding sequences** are transcribed into mRNA which is then translated into protein.
- Most of the genome consists of **non-coding sequences** which are transcribed into either tRNA or rRNA. These are used in protein synthesis but are not translated (Figure 1.20).
- Other non-coding sequences regulate transcription and can work, for example, by switching a gene off so that it is not transcribed.

Key term

Non-coding sequence
A sequence of bases on DNA that does not encode a protein.

Exam tip

In your exam you may be asked to explain the differences between coding and non-coding sequences of DNA. Coding sequences (genes) code for proteins. Non-coding sequences are regulatory sequences that regulate transcription *or* are transcribed into tRNA or rRNA but are not translated.

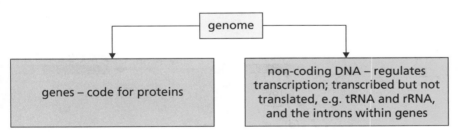

Figure 1.20 Summary of the components of the genome

Personalised medicine

- The nucleotide base sequence can be determined for each individual and analysed to predict the likelihood of the individual developing certain diseases.
- **Pharmacogenetics** is the use of genome information to make informed choices about the drugs and dosages that would be most effective for an individual – this idea is called **personalised medicine**, as shown in Figure 1.21.

Key term

Pharmacogenetics Use of an individual's genome information to select the most effective drugs and dosage in treatment.

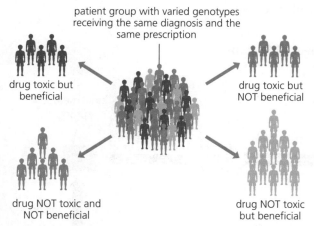

patient group with varied genotypes receiving the same diagnosis and the same prescription

drug toxic but beneficial

drug toxic but NOT beneficial

drug NOT toxic and NOT beneficial

drug NOT toxic but beneficial

Figure 1.21 How a patient group might respond to a drug. Use of individual genome sequence data might allow the prediction of which group a patient is in before the drug is prescribed

Exam tip

Be aware of the potential benefit of personalised medicine to patients for your exam. Type of drug, dose and treatments are tailored to an individual's genome. Different treatments can be designed for different individuals.

Do you know?

1 Describe the structure of the human genome. [5]
2 Give an account of personalised genomics and medicine. [4]

1.6 Metabolic pathways

You need to know

- that metabolic pathways are integrated, controlled pathways of enzyme-catalysed reactions within a cell
- that metabolic pathways are controlled by the presence or absence of particular enzymes
- how the rate of reaction of key enzymes is regulated
- that protein pores, pumps and enzymes are embedded in membranes

Metabolism in cells

- **Metabolism** is the sum total of all the chemical reactions that take place in the cells of a living organism.
- Metabolism can be thought of as a set of interlinked metabolic pathways.

- A metabolic pathway is a series of integrated chemical reactions, each controlled by a different enzyme.
- Metabolic pathways can have **reversible** steps, **irreversible** steps and alternative routes, as shown in Figure 1.22.

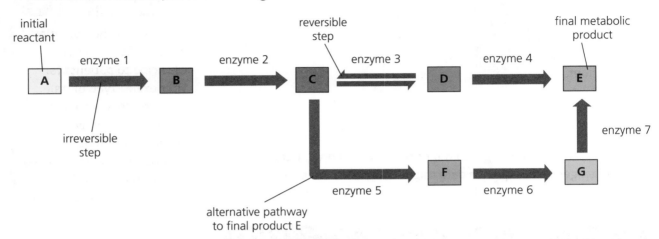

Figure 1.22 A metabolic pathway involving substances A–G and showing irreversible steps, a reversible step involving enzyme 3 and an alternative route by which substance C can be converted to the final metabolic product, E

- Reactions within metabolic pathways can be **anabolic** or **catabolic**, as shown in Figure 1.23.
- Anabolic reactions build up large molecules from small molecules and require/use up energy. Protein synthesis is an example of an anabolic pathway in which energy is used to link amino acids to build up polypeptide molecules.
- Catabolic reactions break down large molecules into smaller molecules and release energy. Cellular respiration is an example of a catabolic pathway in which glucose is broken down into water and carbon dioxide molecules, releasing energy.

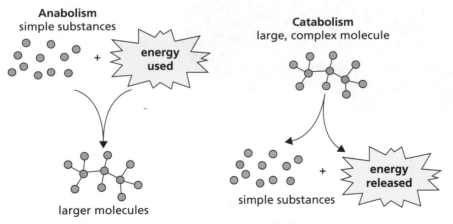

Figure 1.23 Summary of the features of anabolic and catabolic reactions

Synoptic links

Protein synthesis in Key Area 1.3 (page 15) is a good example of anabolism, and respiration in Key Area 1.7 (page 29) is a good example of catabolism.

Control of metabolic pathways

- Metabolic pathways are controlled by the presence or absence of particular enzymes and they are regulated by the rates of reaction of key enzymes in the pathways.
- **Substrate** molecule(s) are acted on by specific enzymes and have a high **affinity** for their **active sites**, and so are very likely to bind with them.
- The **products** made in the reaction have a lower affinity, allowing them to unbind and leave the active sites.
- **Induced fit** is a process that occurs when the active site changes shape to fit the substrate more tightly after the substrate starts to bind with it.
- The energy required to start a chemical reaction is called its **activation energy (E_A)**.
- Enzymes lower the activation energy needed for a reaction to start, as shown in Figure 1.24. This allows the reaction to start more readily and progress at a faster rate than without the enzyme.

Exam tip

In your exam you may be asked to explain the role of genes in the control of metabolic pathways. Remember that genes code for the enzymes that control the metabolic pathway.

Key terms

Affinity Degree of attraction between molecules.

Active site Region on an enzyme molecule where the substrate binds.

Induced fit The process of active-site shape change that binds the substrate molecule more tightly.

Activation energy (E_A) The input of energy required to start a chemical reaction.

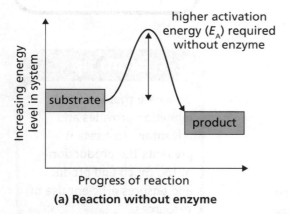

(a) Reaction without enzyme

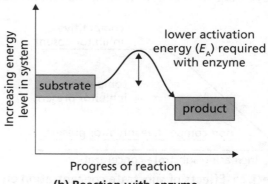

(b) Reaction with enzyme

Figure 1.24 (a) The E_A needed to start the conversion of a substrate to a product; (b) the presence of the specific enzyme lowers the E_A and this allows the reaction to progress at a higher rate. Note that the reaction shown is catabolic because the energy level of the product is less than the substrate, that is, energy has been released

- The concentration of substrate and end product affect the direction and rate of an enzyme reaction. As the substrate concentration increases, the rate of the enzyme reaction increases until all of the active sites are occupied by the substrate, and so the rate of reaction levels off.
- Some metabolic reactions are reversible and the concentration of their substrate(s) relative to the concentration of their product(s) can affect the net direction of the reaction.

Inhibitors

- Inhibitors are substances that can reduce the rate of enzyme-controlled reactions.
- In **competitive inhibition**, the overall shape of the inhibitor molecules resembles that of the normal substrate so they can temporarily bind to occupy active sites on enzyme molecules, preventing binding by the substrate and slowing the reaction rate.
- The effects of competitive inhibition can be reversed by increasing the substrate concentration so that there is a higher chance of a substrate molecule binding, as shown in Figure 1.25.
- **Non-competitive inhibitors** bind at sites on the enzyme molecules which are away from the active site, but their binding permanently changes the shape of the active site, preventing the substrate from ever binding.
- Non-competitive inhibition cannot be reversed by increasing the substrate concentration because the binding of the inhibitor makes a permanent change to an enzyme molecule.
- **Feedback inhibition** occurs when the end product in a metabolic pathway reaches a critical concentration and inhibits an enzyme involved in an earlier stage of the pathway, so reducing or preventing further synthesis of the end product.

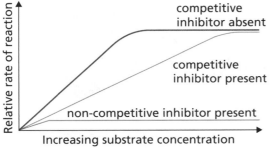

Figure 1.25 Effects of substrate concentration on the rate of an enzyme reaction in the absence and presence of inhibitors. Note that in competitive inhibition, the inhibitor has less effect at higher substrate concentrations, but increasing substrate concentration has no effect on the action of non-competitive inhibitors

Technique

Using substrate concentration and inhibitor concentration to alter reaction rates are techniques you are expected to be familiar with for your exam. Reactions can be controlled in experiments to produce results like those shown in graphs (a) and (b) of Figure 1.26 below.

In graph (a), the concentration of the substrate is increased from 0 to X and the rate of reaction increases because there are plenty of active sites to bind the extra substrate. Increase of substrate between X and Y has no effect on rate because all the active sites of the enzyme are continuously full. In graph (b), increasing the concentration of the inhibitor progressively blocks more and more active sites, stopping binding by substrate molecules and reducing the reaction rate of the enzyme.

(a)

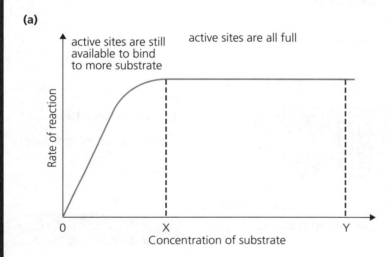

(b)

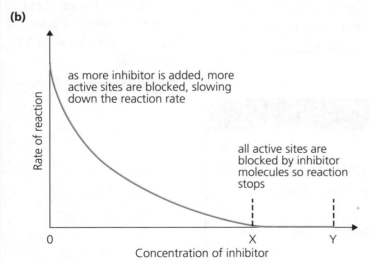

Figure 1.26 The effects of (a) substrate and (b) inhibitor concentrations on the rate of an enzyme-controlled reaction

Exam tip

In your exam you may be asked to describe the difference between a competitive inhibitor and non-competitive inhibitor. The competitive inhibitor molecule has a similar shape to the substrate molecule and binds to the active site. The non-competitive inhibitor molecule binds to a different site on the enzyme and changes the shape of the active site.

Membranes

■ Proteins embedded in **phospholipid** membranes have functions such as pores, pumps or enzymes, as shown in Figure 1.27.

  □ Pores are gaps in the membrane through which certain substances can pass.

  □ Pumps are able to use energy to move substances actively from one side of a membrane to the other.

  □ Membrane-bound enzymes speed up specific reactions.

Exam tip

Remember the mnemonic '**PEP**' to help you describe the roles of proteins embedded in phospholipid membranes – **p**ores, **e**nzymes and **p**umps.

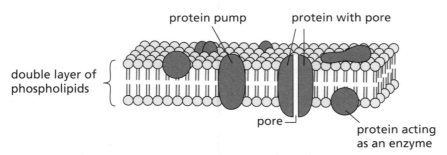

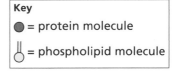

Figure 1.27 **A phospholipid membrane with specialised proteins**

■ An example of a membrane that has metabolic functions is the inner mitochondrial membrane involved in aerobic respiration. Notice that this membrane also has carrier proteins as well as pumps and enzymes.

Synoptic link

You can read more about the inner mitochondrial membrane in Key Area 1.7 (page 29).

Do you know?

1 Write notes on metabolic pathways. [6]

2 Give an account of the mechanism of enzyme action. [5]

3 Give an account of the competitive and non-competitive inhibition of enzymes. [4]

4 Give an account of the structure of membranes in cells and the types of protein that are embedded. [4]

1.7 Cellular respiration

You need to know

■ that ATP is synthesised during cellular respiration
■ the role of ATP in the transfer of energy
■ how to describe the metabolic pathways of glycolysis, the citric acid cycle and the electron transport chain in aerobic respiration

Synthesis and role of ATP

■ ATP is synthesised during aerobic cellular respiration.
■ ATP produced in cellular respiration transfers energy to anabolic processes such as protein synthesis, as shown in Figure 1.28.

Synoptic link

You can read more about protein synthesis in Key Area 1.3 (page 15).

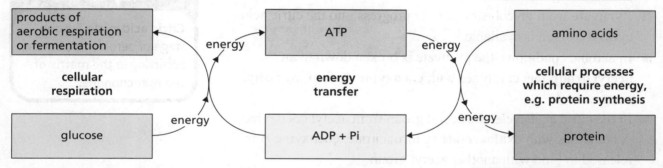

Figure 1.28 Summary of the role of ATP in energy transfer – the energy released from cellular respiration is transferred to the anabolic protein synthesis

Glycolysis

■ **Glycolysis** is the breakdown of glucose to pyruvate in the cytoplasm of cells, as shown in Figure 1.29.
■ Two molecules of ATP are required for the phosphorylation of one glucose molecule and intermediates in the metabolic pathway during the **energy investment** phase of glycolysis.
■ This leads on to the generation of 4 molecules of ATP during the **energy pay-off** stage and results in a net gain of 2 ATP.
■ More ATP is made in glycolysis than is needed to start it.
■ In glycolysis, **dehydrogenase enzymes** remove hydrogen ions and **electrons** from substrates and pass them to the coenzyme **NAD**, forming NADH – this also occurs in the citric acid cycle.

Key terms

Glycolysis The first stage of cellular respiration in the cytoplasm, involving the breakdown of glucose to pyruvate with a pay-off/net gain of 2 ATP.

Dehydrogenase enzymes Remove hydrogen ions and electrons and pass them to the coenzyme NAD, forming NADH.

Electrons Negatively charged particles; these are passed along the electron transport chain, releasing energy that allows hydrogen ions to be pumped across the inner mitochondrial membrane.

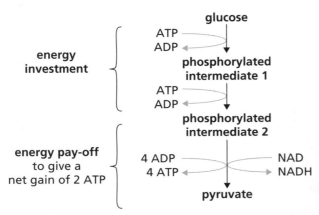

Figure 1.29 Summary of the main features of glycolysis. Note the net gain of 2 ATP and the production of NADH

Citric acid cycle

- The **citric acid cycle** occurs in the matrix of the mitochondria, as shown in Figure 1.30.
- **Pyruvate** from glycolysis can only progress into the citric acid cycle if oxygen is available.
- In aerobic conditions, the pyruvate is broken down to an **acetyl group** that combines with **coenzyme A**, forming acetyl coenzyme A.
- In the citric acid cycle, the acetyl group from acetyl coenzyme A combines with **oxaloacetate** to form citrate; coenzyme A is released to bind with another acetyl group.
- During the series of enzyme-controlled steps of the citric acid cycle, citrate is gradually converted back into oxaloacetate, resulting in the generation of ATP and the release of carbon dioxide as a final metabolic product.
- As in glycolysis, **dehydrogenase enzymes** remove hydrogen ions and electrons from substrates and pass them to the coenzyme NAD, forming NADH.

Exam tip

Make sure that you understand the role of dehydrogenase enzymes and NAD in the citric acid cycle. Remember it is all about hydrogen ions and electrons being removed from substrates and transferred to the carrier NAD. NADH then transports the hydrogen ions and electrons to the electron transport chain (on the inner mitochondrial membrane).

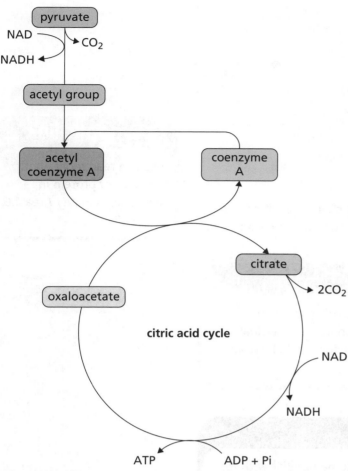

Figure 1.30 Summary of the main steps in the citric acid cycle. Note the production of carbon dioxide, ATP and NADH

The electron transport chain

- The hydrogen ions and electrons carried by NADH are passed to the **electron transport chain**, which is a series of carrier proteins attached to the inner mitochondrial membrane, as shown in Figure 1.31.
- Electrons are passed along the carrier proteins of the electron transport chain, releasing energy as they flow.
- This energy allows hydrogen ions to be pumped across the inner mitochondrial membrane and cause a high concentration on the outside.
- The return flow of the hydrogen ions down the concentration gradient back through the membrane protein **ATP synthase** results in the synthesis of ATP.
- On the inside of the membrane, hydrogen ions and electrons finally combine with oxygen to form water, which is a final metabolic product.

Key term

ATP synthase An enzyme (protein) embedded in the inner membrane of the mitochondria that produces ATP as hydrogen ions flow through it, down a concentration gradient.

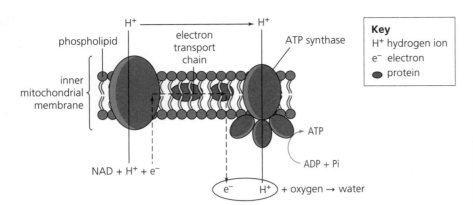

Figure 1.31 Summary of the electron transport chain – significantly more ATP is made at this stage than in glycolysis or the citric acid cycle

Synoptic link

You can read more about the roles of protein in metabolism in Key Area 1.6 (page 23).

Exam tip

In your exam you may be asked to explain the role played by oxygen in the electron transport chain. Oxygen acts as the final electron acceptor and combines with hydrogen ions and electrons to form water.

Technique

A **respirometer** is a device you are expected to be familiar with for your exam. It is designed to measure the rate of respiration in cells, tissues and organisms.

The simplest type of respirometer measures the rate of consumption of oxygen by a whole organism in a closed container, as shown in Figure 1.32.

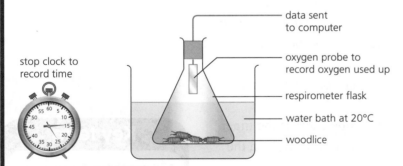

Figure 1.32 A simple respirometer set up to measure respiration rate in woodlice; the oxygen they consume in a set period of time is measured

Technique

You are expected to be familiar with a technique for **measuring metabolic rate**.

Metabolic rate is related to the usage of oxygen or production of carbon dioxide in respiration. High metabolic rate is linked to high volumes of oxygen being consumed and high volumes of carbon dioxide being produced.

Oxygen usage can be measured with an oxygen probe and carbon dioxide production with a carbon dioxide probe. You should be aware that temperature probes that measure heat production by the body are also used to give values for metabolic rate. Figure 1.33 shows a human subject in a chamber in which probes are being used to monitor metabolic rate.

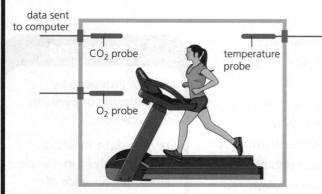

Figure 1.33 **Metabolic rate being monitored in a human subject**

Do you know?

1 Describe the energy investment and energy pay-off phases in glycolysis. [4]

2 Give an account of the citric acid cycle in respiration. [7]

3 Give a description of the electron transport chain in cellular respiration. [6]

4 Describe a respirometer. [2]

1.8 Energy systems in muscle cells

You need to know

- about lactate metabolism
- the structure and function of different types of skeletal muscle fibre

Lactate metabolism

- During vigorous exercise such as sprinting, breathing rates increase but muscle cells still do not receive sufficient oxygen to support the electron transport chain.
- In these conditions, pyruvate produced by glycolysis in muscle cells is converted to **lactate** and an **oxygen debt** builds up.
- The conversion of pyruvate to lactate requires hydrogen from the NADH produced during glycolysis. This conversion regenerates NAD, which is needed to maintain ATP production through glycolysis, as shown in Figure 1.34.

Key terms

Lactate Produced by the fermentation of pyruvate in muscle cells.

Oxygen debt Builds up during glycolysis in muscle cells in the absence of sufficient oxygen and has to be repaid during rest periods after exercise.

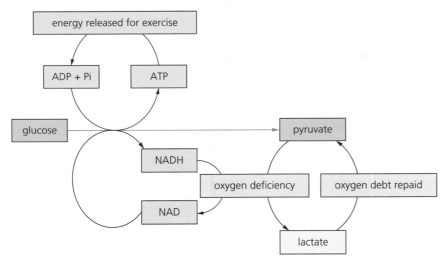

Figure 1.34 Summary of lactate metabolism and repayment of the oxygen debt

- Lactate is toxic and accumulates in affected muscle cells, causing painful muscle fatigue.
- When exercise is complete, heavy breathing continues for a time during which the oxygen debt is repaid.

- During repayment, oxygen is used in aerobic respiration to provide further ATP.
- ATP is required to convert the toxic lactate back to pyruvate and eventually into glucose in the liver, as shown in Figure 1.35.

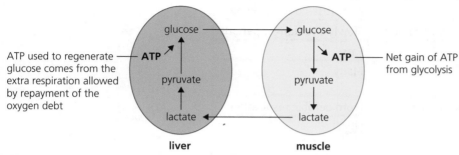

ATP used to regenerate glucose comes from the extra respiration allowed by repayment of the oxygen debt

Net gain of ATP from glycolysis

liver   **muscle**

Figure 1.35 During repayment of the oxygen debt, muscle cells respire aerobically to produce ATP; liver cells use this ATP to convert lactate into pyruvate and then into glucose

Slow-twitch and fast-twitch muscle fibres

- **Slow-twitch** muscle **fibres**:
  - ☐ contract relatively slowly, but can sustain contractions for longer than fast-twitch muscle fibres
  - ☐ are useful for endurance activities such as long-distance running, cycling or cross-country skiing
  - ☐ rely on aerobic respiration to generate ATP and their cells have many mitochondria, a large blood supply and a high concentration of the oxygen-storing protein myoglobin
  - ☐ use fats as their major storage fuel
- **Fast-twitch muscle fibres**:
  - ☐ contract relatively quickly and their action occurs over short periods
  - ☐ are useful for activities that require bursts of energy release, such as sprinting or weightlifting
  - ☐ can generate ATP through glycolysis only and their cells have fewer mitochondria and a lower blood supply compared with slow-twitch muscle fibres
  - ☐ use **glycogen** as their major storage fuel

> **Key term**
>
> **Glycogen** Storage carbohydrate built up from glucose and stored in liver and muscle cells.

Table 1.1 Differences between fast-twitch and slow-twitch muscle fibres

Characteristic	Fast-twitch fibres	Slow-twitch fibres
Activities	Speed and strength, e.g. sprinting and weightlifting	Stamina and endurance, e.g. long-distance running and cross-country skiing
Contractions	Quickly over short periods	Slowly over longer periods
Major storage fuel	Glycogen	Fats
Number of mitochondria	Few	Many
ATP generation	From glycolysis only	From aerobic respiration
Concentration of myoglobin (for oxygen storage)	Low	High
Blood supply	Low	High

- Most human muscle tissue contains a mixture of both slow- and fast-twitch muscle fibres.
- Athletes show distinct patterns of muscle fibres that reflect their sporting activities, as shown in Figure 1.36.

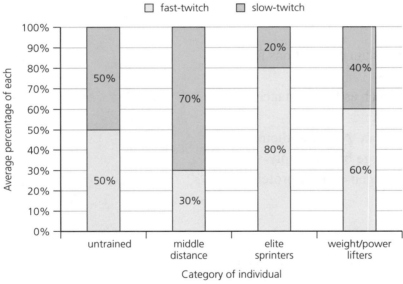

Exam tip

Be clear on muscle fibre types: if the activity involves **fast** movement, the trained athlete develops more **fast-twitch** fibres. Make sure you can relate percentages of different fibre types with distinct types of exercise.

Figure 1.36 Distinct proportions of different muscle fibre types in individuals who have trained for different events. Compare the relative percentages of fast-twitch fibres in sprinters with weightlifters

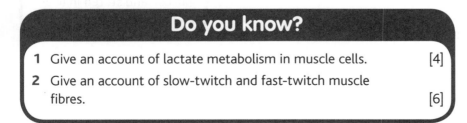

Do you know?

1 Give an account of lactate metabolism in muscle cells. [4]
2 Give an account of slow-twitch and fast-twitch muscle fibres. [6]

Area 1 assessment
Human cells

Answer on separate sheets of paper. Mark your own work at: **hoddereducation.co.uk/needtoknow/answers**

1 The human body contains many specialised cells, all of which have developed from stem cells.

   a Give the term used to describe the process by which a cell develops specialised functions. [1]

   b Describe *one* way in which tissue stem cells differ from embryonic stem cells. [1]

   c Describe how cancer cells form a tumour and explain how secondary tumours arise. [2]

2 DNA is a substance that encodes the genetic information of heredity in a chemical language.

   a Explain what is meant by the following terms as applied to DNA structure:

      i complementary base pairing [1]

      ii antiparallel [1]

   b Name the *two* components of DNA nucleotides that are combined to form the backbone of each strand of DNA. [1]

   c State the structural difference between the 3′ and 5′ end of a DNA strand. [1]

   d DNA is encoded in triplet sequences. Explain what is meant by this. [1]

3 DNA replicates before cell division and copies are passed to daughter cells.

   a Describe the roles of the enzymes DNA polymerase and ligase in DNA replication. [2]

   b The leading strand of DNA is replicated continuously, while the lagging strand can only be replicated in fragments. Explain why the strands are replicated in different ways. [1]

4 The polymerase chain reaction (PCR) is a technique that is carried out in the laboratory and involves cycles of heating and cooling.

   a State the function of PCR. [1]

   b Explain the purpose of different temperatures in the cycles of heating and cooling in PCR. [2]

   c State the number of DNA molecules that would be present after one molecule of DNA has passed through seven thermal cycles of PCR. [1]

   d Describe *one* application of the PCR procedure. [1]

5 Gene expression is the process by which specific genes are activated to produce a required protein.

   a Give an account of the transcription of DNA to form a primary transcript. [2]

   b Name the process by which the coding regions of a primary mRNA transcript are joined together to produce a mature mRNA transcript. [1]

   c Give an account of the translation of mature mRNA at a ribosome. [2]

   d Proteins are chains of amino acids folded into three-dimensional shapes.

      i Name the bonds that hold the amino acids together in sequence. [1]

      ii Name a bond that holds proteins in three-dimensional shapes. [1]

6 Single-gene mutations occur within genes and involve changes to the DNA nucleotide sequence, while other mutations affect the structure of chromosomes.

   a Explain what is meant by a 'mutation'. [1]

   b Describe the effects of the following types of mutation:

      i   missense mutation [1]

      ii nonsense mutation [1]

      iii splice site mutation. [1]

   c Explain the likely effect on the structure of an enzyme of a mutation in which an extra nucleotide is inserted into the gene. [1]

   d Describe a duplication and a translocation chromosome mutation. [2]

7 a Describe what is meant by the 'genome' of an organism. [1]

   b Give *one* function of the non-coding regions of the genome. [1]

   c Give the meaning of the following terms:

      i   sequence data [1]

      ii bioinformatics. [1]

   d Explain how the analysis of individual genomes may lead to personalised medicines. [1]

8 Metabolism is all the chemical reactions that take place in cells.

   a Describe the role of genes in the control of metabolic pathways. [2]

   b Explain what is meant by the induced-fit model of enzyme action. [1]

   c Give *one* difference between anabolic and catabolic pathways. [1]

   d Describe the effect of an increase in substrate concentration on the direction and rate of an enzyme reaction. [1]

9 During cellular respiration, glucose is broken down in a series of enzyme-controlled stages.

   a Describe what happens during the energy investment and energy pay-off phases of glycolysis. [2]

   b State the role of dehydrogenase enzymes in glycolysis and the citric acid cycle. [1]

   c Describe the role of the coenzyme NAD. [1]

   d Explain the role of NADH when cells do not get sufficient oxygen for aerobic respiration. [2]

   e Name the enzyme embedded in the inner membrane of a mitochondrion that is responsible for the regeneration of ATP. [1]

   f Describe the role of the electrons transported to the electron transport chain. [1]

   g State the role of oxygen in the electron transport chain. [1]

Extended response

10 Give an account of skeletal muscle cells under the following headings:

   a lactate metabolism [4]

   b slow-twitch muscle fibres [3]

   c fast-twitch muscle fibres [3]

2 Physiology and health

2.1 Gamete production and fertilisation

You need to know
- how gametes are produced in the testes
- how gametes are produced in the ovaries
- about fertilisation

Gamete production

- Gametes are produced from **germline** cells by **meiosis**.
- Sperm are produced in the **seminiferous tubules** in the testes.
- The **interstitial** cells of the testes produce the hormone **testosterone**, which stimulates sperm production.
- The **prostate gland** and **seminal vesicles** secrete fluids which are added to the sperm and help to maintain their motility and viability by giving them energy to move.

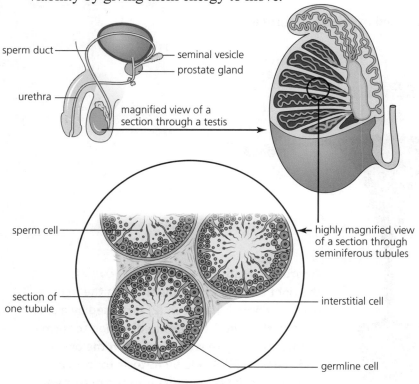

Figure 2.1 Details of male gamete production

2 Physiology and health

- The ovaries contain immature ova (egg cells) in various stages of development.
- Each ovum is surrounded by a **follicle** that protects the developing ovum and secretes the ovarian hormones called **oestrogen** and **progesterone**.

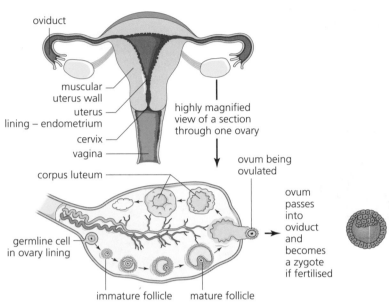

Figure 2.2 Details of female gamete production

Fertilisation

- Once ova become mature, they are released into the oviduct in the process of ovulation.
- In the oviducts, an ovum can be fertilised by sperm to form a zygote, as shown in Figure 2.3.

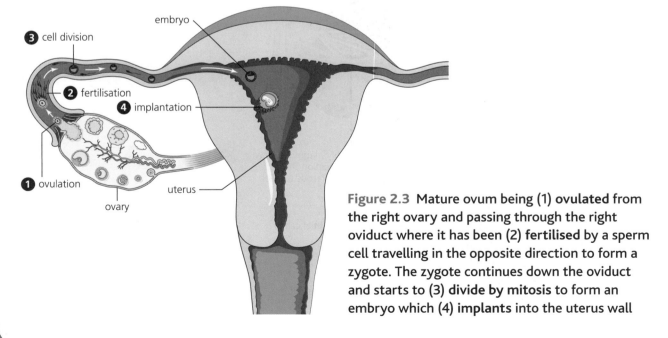

Figure 2.3 Mature ovum being (1) **ovulated** from the right ovary and passing through the right oviduct where it has been (2) **fertilised** by a sperm cell travelling in the opposite direction to form a zygote. The zygote continues down the oviduct and starts to (3) **divide by mitosis** to form an embryo which (4) **implants** into the uterus wall

Do you know?

1 Give an account of gamete production in the testes and the roles of the prostate gland and seminal vesicles in fertility. [6]
2 Give an account of ovulation and fertilisation in females. [3]

2.2 Hormonal control of reproduction

You need to know

- how hormones influence puberty
- how hormones control sperm production
- how hormones control the menstrual cycle

Role of hormones in reproduction

- Hormones are chemical messengers produced by the **endocrine glands**.
- They are released directly into the bloodstream and travel to their target tissue or organ where they have their effect.
- Hormones control the onset of **puberty**, sperm production in males and the **menstrual cycle** in females.
- At puberty, the **hypothalamus** in the brain secretes a releaser hormone that targets the pituitary gland at the base of the brain.
- The pituitary gland is stimulated to release a hormone called **follicle stimulating hormone (FSH)** in both males and females and:
  - □ **luteinising hormone (LH)** in women
  - □ **interstitial cell stimulating hormone (ICSH)** in men
- These hormones trigger the onset of puberty.

Hormones in males

- In males, FSH promotes sperm production in the seminiferous tubules of the testes.
- ICSH stimulates the interstitial cells in the testes to produce the male sex hormone testosterone.

- Testosterone stimulates sperm production and activates the prostate gland and seminal vesicles to produce their fluid secretions (Figure 2.4).

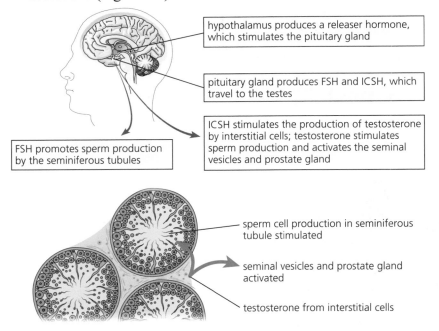

Figure 2.4 **Puberty and the onset of sperm production in males. Note the names and functions of the hormones**

- Overproduction of testosterone is prevented by a **negative feedback mechanism**.
- High testosterone levels inhibit the secretion of FSH and ICSH from the **pituitary gland**, resulting in a decrease in the production of testosterone by the interstitial cells, as shown in Figure 2.5.

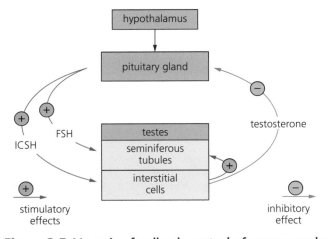

Figure 2.5 **Negative feedback control of sperm production in males. Note that testosterone feeds back to inhibit the pituitary gland**

Hormones in females

- The menstrual cycle takes approximately 28 days, with the first day of menstruation regarded as day one of the cycle.

- The pituitary hormones, follicle stimulating hormone (FSH) and luteinising hormone (LH), and the ovarian hormones, oestrogen and progesterone, are associated with the menstrual, cycle as shown in Figure 2.6.

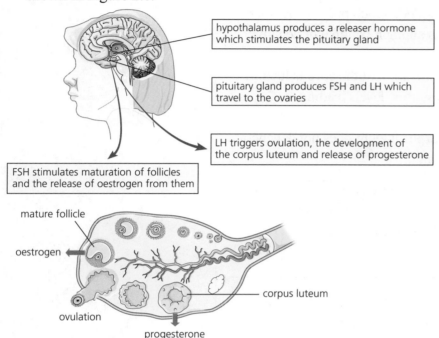

Figure 2.6 Puberty and the onset of ovum production in females. Note the names and functions of the hormones

Follicular phase

- In the **follicular phase** (the first half of the cycle), FSH stimulates the development and maturation of a follicle surrounding the ovum and the production of the sex hormone oestrogen by the follicle (Figure 2.7).
- Oestrogen stimulates the repair and vascularisation of the **endometrium**, thickening it and preparing it for implantation.
- Oestrogen also affects the consistency of the **cervical mucus**, making it more easily penetrated by sperm.
- Peak levels of oestrogen stimulate a surge in the secretion of LH by the pituitary gland.

Luteal phase

- In the **luteal phase** (the second half of the cycle), the surge in LH triggers ovulation.
- **Ovulation** is the release of an egg (ovum) from a follicle in the ovary. It usually occurs around the midpoint of the menstrual cycle.
- LH also triggers the development of the **corpus luteum** from the follicle and stimulates it to secrete the sex hormone progesterone.

- Progesterone promotes further development and vascularisation of the endometrium, preparing it for implantation if fertilisation occurs.
- High levels of oestrogen and progesterone inhibit the secretion of FSH and LH by the pituitary gland, which prevents further follicles from developing. This is another example of negative feedback control.
- The lack of LH leads to the degeneration of the corpus luteum, with a subsequent drop in progesterone levels leading to menstruation.
- If fertilisation does occur, the corpus luteum does not degenerate and progesterone levels remain high.

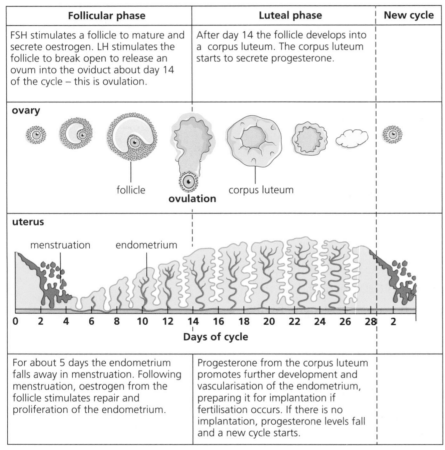

Follicular phase	Luteal phase	New cycle
FSH stimulates a follicle to mature and secrete oestrogen. LH stimulates the follicle to break open to release an ovum into the oviduct about day 14 of the cycle – this is ovulation.	After day 14 the follicle develops into a corpus luteum. The corpus luteum starts to secrete progesterone.	
For about 5 days the endometrium falls away in menstruation. Following menstruation, oestrogen from the follicle stimulates repair and proliferation of the endometrium.	Progesterone from the corpus luteum promotes further development and vascularisation of the endometrium, preparing it for implantation if fertilisation occurs. If there is no implantation, progesterone levels fall and a new cycle starts.	

Figure 2.7 Events of the menstrual cycle in females. Note that each cycle occurs in two phases

Do you know?

1 Give an account of the onset of puberty.   [5]
2 Give an account of the negative feedback control of sperm production.   [3]
3 Give an account of the role of pituitary and ovarian hormones in the menstrual cycle.   [6]

2.3 Biology of controlling fertility

You need to know

- that women have cyclical fertility and men have continuous fertility
- that infertility treatments and contraception are based on the biology of fertility
- the main methods used to treat infertility
- the physical and chemical methods of contraception

Fertility

- Men continually produce sperm in their testes, so show continuous fertility.
- Continuous fertility is due to the relatively constant levels of pituitary hormones.
- Women are only fertile for a few days during each menstrual cycle and so have cyclical fertility, leading to a fertile period, as shown in Figure 2.8.

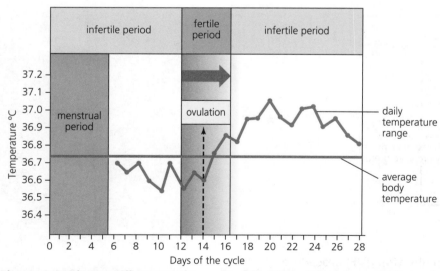

Figure 2.8 Chart to illustrate the cyclical fertility of females as reflected in body temperature

■ The time with the highest likelihood of pregnancy resulting from sexual intercourse is the woman's fertile period, from a few days before until 1–2 days after ovulation.

■ Time of ovulation can be estimated based upon a slight rise in body temperature on the day of ovulation and the thinning of **cervical mucus**.

■ A woman's body temperature rises by around 0.5°C after ovulation and her cervical mucus becomes thin and watery.

Infertility

■ Female **infertility** may be due to failure to ovulate, which is usually the result of a hormone imbalance.

   ☐ Ovulatory or fertility drugs can be used to stimulate ovulation.

   ☐ Some ovulatory drugs work by preventing the negative feedback effect of oestrogen on FSH secretion.

   ☐ Other ovulatory drugs mimic the action of FSH and LH.

   ☐ Ovulatory drugs can cause superovulation, which can result in multiple births or be used to collect ova for **in vitro fertilisation (IVF)** programmes.

■ IVF involves the surgical removal of eggs from ovaries after hormonal stimulation, mixing with sperm to achieve fertilisation in a culture dish, incubating zygotes to form embryos of at least eight cells and uterine implantation, as shown in Figure 2.9.

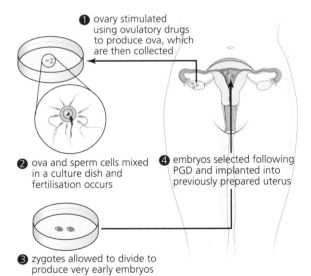

❶ ovary stimulated using ovulatory drugs to produce ova, which are then collected

❷ ova and sperm cells mixed in a culture dish and fertilisation occurs

❹ embryos selected following PGD and implanted into previously prepared uterus

❸ zygotes allowed to divide to produce very early embryos

Figure 2.9 Stages in the in vitro fertilisation (IVF) procedure

■ IVF is used in conjunction with **pre-implantation genetic diagnosis (PGD)** to identify single-gene disorders and chromosomal abnormalities.

- **Artificial insemination (AI)** is a medical treatment in which semen is inserted into the female reproductive tract using a syringe and without intercourse having taken place.
  - ☐ Artificial insemination is particularly useful where the male has a low sperm count.
  - ☐ Several samples of semen are collected and combined over a period of time.
  - ☐ If a male partner is infertile (sterile), a donor may be used to provide semen for artificial insemination.
- If mature sperm are defective or very low in number, **intracytoplasmic sperm injection (ICSI)** can be used.
- This procedure involves the head of a sperm being drawn into a needle and injected directly into the egg to achieve fertilisation, as shown in Figure 2.10.

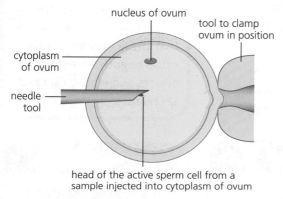

nucleus of ovum

tool to clamp ovum in position

cytoplasm of ovum

needle tool

head of the active sperm cell from a sample injected into cytoplasm of ovum

Figure 2.10 Intracytoplasmic sperm injection (ICSI)

Contraception

- Contraception is the intentional prevention of pregnancy (conception) by natural or artificial methods.
- Contraception includes both physical and chemical methods.
- Physical methods of contraception have a biological basis and include barrier methods, **intra-uterine devices (IUDs)** (shown in Figure 2.11) and sterilisation procedures (shown in Figure 2.12).

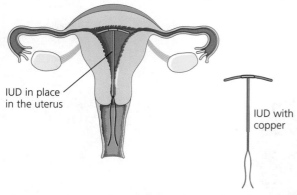

IUD in place in the uterus

IUD with copper

Figure 2.11 An intra-uterine device (IUD): a small structure that is often T-shaped with metallic copper parts fitted into the uterus to prevent the implantation of an embryo in the endometrium

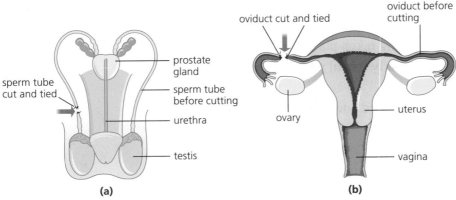

Figure 2.12 Sterilisation by (a) vasectomy in males, which involves cutting and closing the sperm tube (vas deferens) of each testis, and (b) tubal ligation in females, which involves cutting or closing each oviduct (fallopian tube)

- The **oral contraceptive pill** is a chemical method of contraception.
- It contains a combination of synthetic oestrogen and progesterone that mimics negative feedback, preventing the release of FSH and LH from the pituitary gland.
- The **progesterone-only (mini) pill** causes thickening of the cervical mucus, preventing the entry of sperm and fertilisation.
- The **morning-after pill** prevents ovulation or implantation.

Exam tip

In your exam you may be asked to explain the effect of hormones in the contraceptive pill on the menstrual cycle. Remember that these hormones can cause negative feedback effects and inhibit FSH and LH release.

Do you know?

1 Explain the biological basis for the stimulation of ovulation by ovulatory drugs. [5]
2 Describe the process of in vitro fertilisation (IVF). [4]
3 Discuss procedures that can be used to treat male infertility. [5]
4 Describe the biological action of the combined oral contraceptive pill. [3]

Synoptic link

You can read more about negative feedback in the menstrual cycle in Key Area 2.2 (page 41).

2.4 Antenatal and postnatal screening

You need to know

- about antenatal screening
- how to analyse patterns of inheritance in genetic screening and counselling
- about postnatal screening

Monitoring techniques

- **Antenatal or prenatal screening** involves testing for diseases or conditions in a fetus or embryo before it is born.
- Antenatal screening identifies the risk of a disorder so that further tests and a prenatal diagnosis can be offered.
- Common antenatal testing procedures include ultrasound scanning, **amniocentesis** and **chorionic villus sampling (CVS)**.

Scanning

- An **ultrasound scanner** is used to produce an ultrasound image on a computer screen.
- Pregnant women are given two ultrasound scans:
  - **Dating scans** that determine pregnancy stage and due date are used with tests for marker chemicals which vary during pregnancy. A dating scan takes place between 8 and 14 weeks.
  - **Anomaly scans** may detect serious physical abnormalities in the fetus. An anomaly scan takes place between 18 and 20 weeks.

Blood and urine tests

- Routine blood and urine tests are carried out throughout pregnancy to monitor the concentrations of **marker chemicals**.
- Marker chemicals are produced during normal physiological changes that take place during pregnancy.
- Measuring a marker chemical at the wrong time could lead to a **false positive result**.
- An atypical concentration of a chemical can lead to **diagnostic testing** to determine if the fetus has a medical condition.

Diagnostic tests

- Diagnostic tests include amniocentesis and chorionic villus sampling (CVS) from the placenta, as shown in Figure 2.13.
- Amniocentesis and CVS allow a prenatal diagnosis to be made and can confirm the presence of conditions such as Down's syndrome.
- Amniocentesis has a small risk of miscarriage.
- CVS can be carried out earlier in pregnancy than amniocentesis, though it has a higher risk of miscarriage.
- Cells from an amniocentesis sample or CVS can be cultured to obtain sufficient cells to produce a **karyotype** to diagnose a range of conditions.
- A karyotype shows the **chromosome complement** of an individual, that is, their chromosomes arranged as homologous pairs.

> **Exam tip**
>
> Be aware of the difference between a screening test and a diagnostic test. A screening test is carried out to identify the risk of a disorder and a diagnostic test is used to confirm the presence of a disorder or condition.

> **Key terms**
>
> **Marker chemicals** Substances produced during pregnancy which are tested for alongside scans.
>
> **False positive result** Error in reporting in which a test result indicates presence of a condition which is actually absent.
>
> **Diagnostic tests** Tests that are used to confirm if the fetus has a medical condition.

> **Exam tip**
>
> In your exam you may be asked to suggest an advantage of using CVS rather than amniocentesis during antenatal screening. Remember that CVS can be carried out earlier in pregnancy.

■ In deciding to proceed with these tests, the element of risk will be assessed, as will the decisions the individuals concerned are likely to make if a test is positive.

(a) amniocentesis

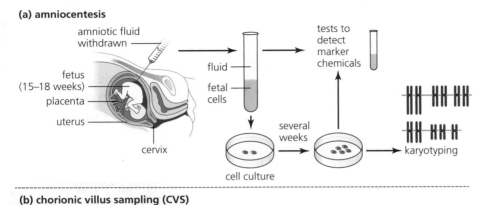

(b) chorionic villus sampling (CVS)

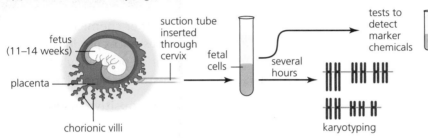

Figure 2.13 Stages of (a) amniocentesis and (b) chorionic villus sampling (CVS)

Patterns in genetic inheritance

■ **Family trees (pedigree charts)** are compiled and used to analyse patterns of inheritance in genetic screening and counselling.
■ Family trees are constructed to provide information and advice in situations where there is the possibility of passing on a genetic disorder to potential offspring.
■ They can be used to analyse patterns of inheritance involving **autosomal recessive, autosomal dominant, autosomal incomplete dominance** and **sex-linked recessive** single-gene disorders.
■ Alleles are forms of the same gene:
  □ Homozygous individuals have two copies of the same allele.
  □ Heterozygous individuals have copies of two different alleles.
■ The chromosome complement of an individual contains 22 pairs of autosomes and one pair of sex chromosomes.

Key terms

Autosomal recessive Allele on chromosomes 1–22; expressed in phenotype only if the genotype is homozygous for the recessive allele.

Autosomal dominant Allele on chromosomes 1–22; always expressed in phenotype.

Autosomal incomplete dominance When an allele is not completely masked by a dominant allele, thus affecting an individual's phenotype.

Sex-linked recessive Recessive allele carried on the X chromosome.

Autosomal inheritance

Autosomal recessive

■ An autosomal recessive disorder, such as cystic fibrosis (CF), is expressed relatively rarely in the offspring.
■ Heterozygous individuals carry the recessive allele but it is not expressed.

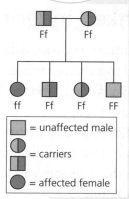

■ It affects males and females equally and may appear to skip generations as it can be passed from two unaffected heterozygous carrier parents to offspring, as shown in Figure 2.14.

Autosomal dominant

■ An autosomal dominant disorder, such as Huntington's disease (HD), shows up in every generation.
■ It affects males and females equally.
■ Heterozygous individuals are affected, as shown in Figure 2.15.

Figure 2.14 Family tree showing the inheritance of the autosomal recessive allele that can cause cystic fibrosis (CF)

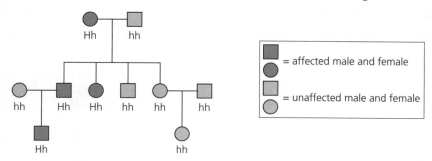

Figure 2.15 Family tree showing the inheritance of the autosomal dominant allele that can cause Huntington's disease (HD)

Autosomal incomplete dominance

■ In examples of autosomal incomplete dominance, such as sickle cell disease (SCD), the fully expressed phenotype of the condition is only seen in homozygous individuals and a severe form of the disorder is rare.
■ The partly expressed phenotype, which gives a less severe form of the condition, is more common because heterozygous individuals can display it.
■ Males and females are affected equally, as shown in Figure 2.16.

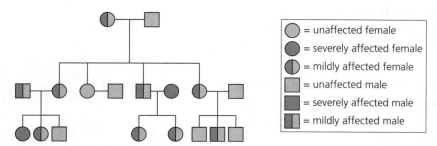

Figure 2.16 Family tree showing the inheritance of the autosomal incompletely dominant allele that can cause sickle cell disease (SCD)

Sex-linked inheritance

- In sex-linked recessive disorders, such as haemophilia, males are affected more than females.
- The allele is carried on the X chromosome, of which males have only a single copy.
- Male offspring receive their X chromosome from their mother.
- Fathers cannot pass the condition on to their sons because they pass on the Y chromosome, which does not carry an allele for the condition.
- Female offspring can only be affected if their father has the condition and their mother is either a carrier or also affected, as shown in Figure 2.17.

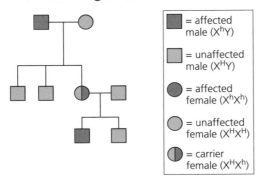

Figure 2.17 Family tree showing the inheritance of the sex-linked recessive allele that can cause haemophilia. Note that the alleles are shown as superscripts on the X chromosome, i.e. X^H and X^h

Exam tip

In your exam you may be asked to explain the different types of inheritable disorders:

- An autosomal recessive disorder is when the allele responsible for the condition is on chromosomes 1–22. It is only expressed in the phenotype if the genotype is homozygous for the recessive allele.
- An autosomal dominant disorder is when the allele responsible for the condition is on chromosomes 1–22. It is always expressed in the phenotype.
- Incomplete dominance is when an allele is not completely masked by a dominant allele and so partly affects an individual's phenotype.
- A sex-linked recessive disorder is when a recessive allele responsible for a disorder is carried on the X chromosome.

Postnatal screening

- Postnatal screening involves health checks that are carried out after the birth of the baby.
- These are aimed at detecting certain conditions or abnormalities.
- Postnatal diagnostic testing is used to detect metabolic disorders such as phenylketonuria (PKU).

Phenylketonuria

- PKU is an inborn error of metabolism caused by an autosomal recessive genetic disorder.
- In PKU, a substitution mutation means that the enzyme that converts the amino acid phenylalanine to tyrosine is non-functional.
- The build-up of phenylalanine, which is converted to a toxic product, affects normal brain development.
- If PKU is not detected soon after birth, the baby's mental development can be affected.
- Individuals with high levels of phenylalanine are placed on a restricted diet which lacks that amino acid.

Do you know?

1 Write notes on the advantages and disadvantages of amniocentesis and chorionic villus sampling as diagnostic techniques. [4]

2 Give an account of the use of family trees (pedigree charts) in genetic screening and counselling. [4]

2.5 Structure and function of arteries, capillaries and veins

You need to know
- the features of blood circulation
- the structure and function of arteries, capillaries and veins
- about the exchange of materials between tissue fluid and cells through pressure filtration and the role of lymphatic vessels

Blood circulation

- Blood circulates from the heart through the **arteries**, to the **capillaries**, then to the **veins** and back to the heart, as shown in Figure 2.18.
- There is a decrease in blood pressure as blood moves away from the heart.

Key terms

Artery A blood vessel that carries blood away from the heart.

Capillary A narrow, thin-walled blood vessel that exchanges materials with the tissues.

Vein A blood vessel with valves that transports blood back to the heart.

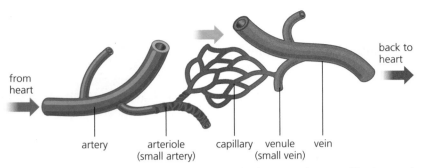

Figure 2.18 Circulation of blood through the arteries, capillaries and veins. Note that small arteries are sometimes known as arterioles

Blood vessels

- Blood vessels are tubes with walls composed of different tissues, depending on the function of the vessel.
- The central space or cavity in blood vessels is called the **lumen**, which is lined with a layer of cells called the **endothelium**.
- The endothelium is surrounded by layers of tissue that differ between arteries, capillaries and veins.

Arteries

- Arteries carry blood away from the heart.
- Blood is pumped through arteries at a high pressure.
- Arteries have an outer layer of **connective tissue** containing elastic fibres, and a thick middle layer containing smooth muscle with more elastic fibres, as shown in Figure 2.19(a).
- The thick elastic walls of the arteries stretch and recoil to accommodate the surge of blood after each contraction of the heart.
- The smooth muscle in the walls of arterioles can contract or relax, causing **vasoconstriction** and **vasodilation** to control blood flow.
- This ability to vasoconstrict or vasodilate allows the changing demands of the body's tissues to be met.
- During exercise the arterioles supplying the muscles vasodilate, which increases the blood flow, and the arterioles supplying the abdominal organs vasoconstrict, which reduces the blood flow to them.

Capillaries

- Capillary walls are only one cell thick.
- This allows quick and efficient exchange of substances with tissues, as shown in Figure 2.19(b).

Veins

- Veins carry blood back towards the heart.
- They have an outer layer of connective tissue containing elastic fibres, but a much thinner muscular wall than arteries, as shown in Figure 2.19(c).
- The central lumen of a vein is wider than that of an artery of equal diameter.
- **Valves** are present in veins to prevent the backflow of blood.
- Valves are needed because blood flows back to the heart at low pressure and generally against the force of gravity.

Exam tip

Make sure you are able to describe the structural differences between an artery and a vein. Veins have valves, a relatively wide lumen and a thinner muscular wall – you find the opposite in arteries.

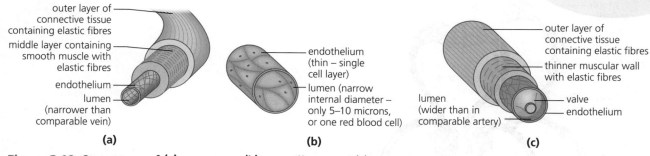

Figure 2.19 Structures of (a) an artery, (b) a capillary and (c) a vein

Exchange of materials and the role of lymphatic vessels

- **Pressure filtration** causes plasma to pass through capillary walls into the tissue fluid that surrounds cells, as shown in Figure 2.20.

Key term

Pressure filtration Passage of molecules through membranes under pressure.

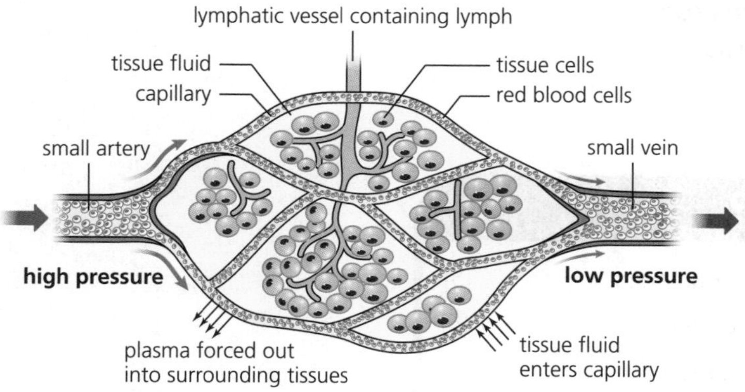

Figure 2.20 Fluids in a capillary bed: plasma, tissue fluid and lymph. Plasma is squeezed out of the capillaries and bathes the cells. Much of this tissue fluid returns to the capillaries directly, but the excess is drained off by the lymph vessels

- Tissue fluid and blood plasma are similar in composition, except that plasma contains plasma proteins, which are too large to be filtered through the capillary walls and so do not appear in the tissue fluid.

Exam tip

In your exam you may be asked to explain the process of pressure filtration in capillary beds. The blood in the arterioles is at a higher pressure than in the capillary bed, which results in pressure filtration, causing the formation of tissue fluid.

- Tissue fluid contains glucose, oxygen and dissolved substances which supply the tissues with all their requirements.
- Useful molecules such as glucose and oxygen diffuse into cells, and carbon dioxide and other metabolic waste substances diffuse out of the cells and into the tissue fluid to be excreted.
- Much of the tissue fluid re-enters the capillaries and returns to the blood.
- The **lymphatic vessels** absorb the remaining excess tissue fluid and return it as **lymph** to the circulatory system, as shown in Figure 2.21.

Key terms

Lymphatic vessels Tiny vessels in which lymph circulates around the body.

Lymph Tissue fluid collected into lymph vessels, which circulates around the body.

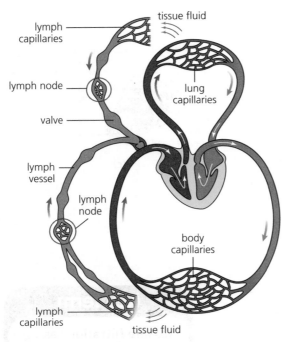

Figure 2.21 The relationship between the circulatory system and the lymphatic system; lymph returns excess tissue fluid to the circulatory system

Do you know?

1 Write notes on the structure and function of arteries. [5]
2 Write notes on the structure and function of veins. [5]
3 Describe the relationship between plasma, tissue fluid and lymph. [3]

2.6 Structure and function of the heart

You need to know

- the definition of 'cardiac output' and its calculation
- about the cardiac cycle
- the structure and function of the cardiac conducting system
- how blood pressure changes in the aorta during the cardiac cycle

Cardiac output

- The heart has four chambers (right atrium, right ventricle, left atrium and left ventricle) and works as a double pump with two separate sides.
- The right side collects deoxygenated blood from the body and pumps it to the lungs to collect oxygen.
- The left side collects oxygenated blood from the lungs and pumps it to the body.

Exam tip

Make sure you revise the meaning of the term 'cardiac output' and how it is calculated.

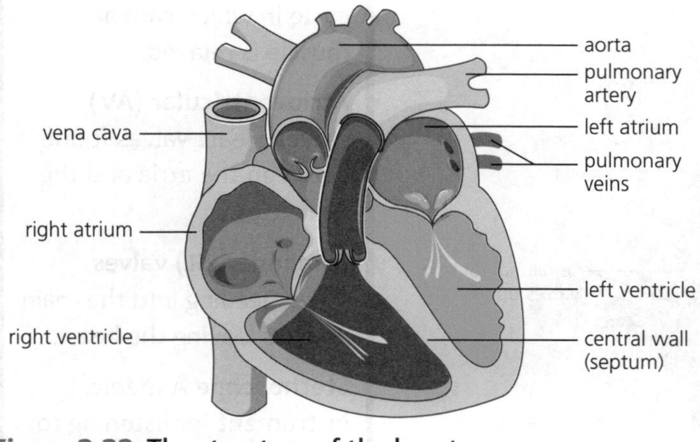

Figure 2.22 The structure of the heart

- The volume of blood pumped through each ventricle per minute is the **cardiac output**.
- Cardiac output is determined by heart rate and **stroke volume** (CO = HR × SV).
- The left and right ventricles pump equal volumes of blood through the aorta and pulmonary artery during one heartbeat.

Key term

Stroke volume The volume of blood expelled from the left ventricle during one cardiac cycle.

The cardiac cycle

- The **cardiac cycle** is the pattern of contraction (**systole**) and relaxation (**diastole**) of the heart muscle in one complete heartbeat, as shown in Figure 2.23.
- During diastole, blood returning to the atria flows into the ventricles through the open **atrioventricular (AV) valves**.
- Atrial systole transfers the remainder of the blood in the atria through the AV valves to the ventricles.
- Ventricular systole closes the AV valves and pumps the blood out through the **semilunar (SL) valves** to the aorta and pulmonary artery.
- In ventricular diastole, the higher pressure in the arteries outside the heart forces the SL valves to close.
- The opening and closing of the AV and SL valves are responsible for the heart sounds ('lubb-dupp') that can be heard with a **stethoscope**.
- The 'lubb' is caused by the closing of the two AV valves, and the 'dupp' by the closing of the two SL valves.

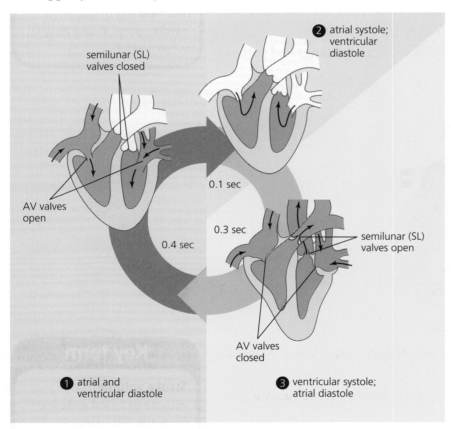

Figure 2.23 The cardiac cycle showing diastole and systole, valve positions and the average duration of each stage

Exam tip

In your exam you may be asked to describe the state of the atrial and ventricular muscles during the diastolic stage of the cardiac cycle. Remember that during diastole, the muscles of the atria and ventricles are relaxed.

Exam tip

Remember that the function of the semilunar valves is to prevent backflow of blood into the ventricles.

Key terms

Systole Part of the cardiac cycle in which cardiac muscle is contracted.

Diastole Part of the cardiac cycle in which cardiac muscle is relaxed.

Atrioventricular (AV) valves Heart valves found between the atria and the ventricles.

Semilunar (SL) valves Valves leading into the main arteries leaving the heart.

Stethoscope A medical instrument for listening to the action of someone's heart or breathing.

The cardiac conducting system

- The heartbeat originates in the heart itself.
- The auto-rhythmic cells of the **sino-atrial node (SAN)** or **pacemaker**, located in the wall of the right atrium, set the rate at which the heart contracts.
- The timing of cardiac muscle cell contraction is controlled by impulses from the SAN spreading through the atria during atrial systole.
- Impulses then travel to the **atrioventricular node (AVN)**, located in the centre of the heart.
- Impulses from the AVN then travel down fibres in the central wall of the heart and up through the walls of the ventricles, causing ventricular systole, as shown in Figure 2.24.
- Impulses in the heart generate currents that can be detected by an **electrocardiogram (ECG)**.
- There are three distinct waves obtained in a normal ECG trace:
  - ☐ The P wave is the wave of electrical impulses spreading over the atria from the SAN.
  - ☐ The QRS complex is when the electrical impulses pass through the ventricles.
  - ☐ The T wave is the electrical recovery of the ventricles at the end of ventricular systole.

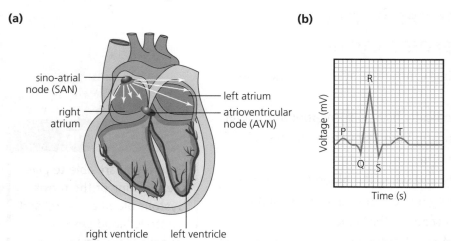

(a)   (b)

Figure 2.24 (a) The cardiac conducting system showing impulses from the SAN being passed across the atria and to the AV node where they can be directed down and through the walls of the ventricles; (b) an ECG trace of one heartbeat

- The medulla in the brain regulates the rate of the SAN through the **antagonistic action** of the **autonomic nervous system (ANS)**.
- The antagonistic action acts like an accelerator and a brake, with one acting against the other:

☐ A **sympathetic nerve** releases **noradrenaline** which increases the heart rate like an accelerator.

☐ A **parasympathetic nerve** releases **acetylcholine** which decreases the heart rate like a brake, as shown in Figure 2.25.

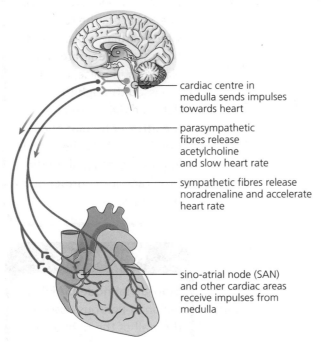

cardiac centre in medulla sends impulses towards heart

parasympathetic fibres release acetylcholine and slow heart rate

sympathetic fibres release noradrenaline and accelerate heart rate

sino-atrial node (SAN) and other cardiac areas receive impulses from medulla

Figure 2.25 The effect of the autonomic nervous system on heart rate

Key terms

Sympathetic nerve Nerve fibre that stimulates an increase in heart rate; part of the ANS.

Parasympathetic nerve Nerve fibre that results in a decrease in heart rate; part of the ANS.

Stroke Life-threatening condition that occurs when blood supply to part of the brain is cut off.

Synoptic link

You can read more about the autonomic nervous system (ANS) in Key Area 3.1 (page 72).

Blood pressure changes in the aorta during the cardiac cycle

- Blood pressure increases during ventricular systole and decreases during diastole.
- Blood pressure is measured medically using a **sphygmomanometer**:
  ☐ A rubber cuff is inflated to stop blood flow in an artery of the arm.
  ☐ It is then deflated gradually and when the blood starts to flow it is detected by a pulse being heard through a stethoscope at **systolic pressure**, which is read off the scale.
  ☐ As the blood flows freely through the artery, the pulse sound is no longer detected at **diastolic pressure** (Figure 2.26).
  ☐ A typical blood pressure reading for a young adult is 120/80 mm Hg.
- **Hypertension** is a condition in which an individual has persistently high blood pressure, a major risk factor for many diseases and conditions, including coronary heart disease (CHD) and **strokes**.

Exam tip

Ensure you are able to give the meaning of the terms 'systolic blood pressure' and 'diastolic blood pressure'. Systolic blood pressure is the pressure that blood exerts on the arteries while the heart is contracting. Diastolic blood pressure is the pressure the blood exerts on the arteries while the heart is relaxed.

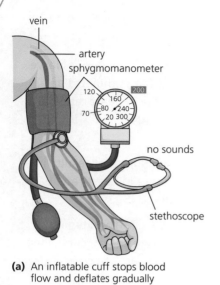

(a) An inflatable cuff stops blood flow and deflates gradually

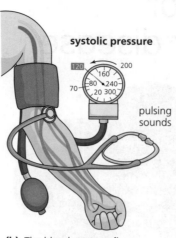

(b) The blood starts to flow (detected by a pulse) at systolic pressure

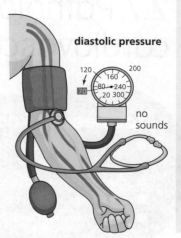

(c) The blood flows freely through the artery (and a pulse is not detected) at diastolic pressure

Figure 2.26 Measurement of blood pressure using a sphygmomanometer; 120 mm Hg is the systolic pressure and in this case 70 mm Hg is the diastolic pressure

Technique

Measuring pulse rates and blood pressures are techniques you are expected to be familiar with for your exam.

Make sure you know that pulse rates are really measures of heartbeats per minute (bpm) and are typically taken where an artery comes close to the skin surface, such as at the wrist. Blood pressure is measured using a sphygmomanometer. A typical reading for a young person is 120/80 – the systolic pressure is 120 mm of Hg (mercury) and the diastolic pressure is 80 mm of Hg (mercury).

Synoptic link

You can read more about blood pressure in Key Area 2.7 (page 62).

Do you know?

1 Write notes on the antagonistic action of the ANS in the control of the cardiac cycle. [4]
2 Give an account of the conducting system of the heart. [6]

2.7 Pathology of cardiovascular disease

You need to know
- the process of atherosclerosis and its effect on arteries and blood pressure
- the formation and possible effects of a thrombosis
- the causes and effects of peripheral vascular disorders
- how cholesterol levels in the body are controlled

Atherosclerosis

- **Atherosclerosis** is the accumulation of fatty material that forms an **atheroma** or **plaque** beneath the endothelium (inner lining) of an artery wall (see Figure 2.27).
- The fatty material consists mainly of **cholesterol**, fibrous material and calcium salts.
- As an atheroma grows, the artery thickens and loses its elasticity. The diameter of the artery lumen becomes narrowed, which restricts blood flow and results in increased blood pressure.
- Atherosclerosis is the root cause of various **cardiovascular diseases (CVD)** including **angina**, **heart attack**, stroke and **peripheral vascular disease**.

Figure 2.27 **Formation of an atheroma beneath the endothelium of an artery wall**

Key terms

Atheroma Swelling on the inner wall of an artery made up of fatty material and connective tissue.

Cholesterol Lipid molecule needed for cell membranes and in synthesising steroid hormones.

Angina Chest pain that occurs when blood supply to the heart muscle is restricted.

Heart attack Serious medical emergency in which blood supply to the heart muscle is blocked; also known as a myocardial infarction (MI).

Peripheral vascular disease Condition caused by blockage to arteries other than the coronary arteries, the aorta or those in the brain.

Exam tip

Remember the effects that atherosclerosis has on the structure of an artery: the lumen diameter is reduced and it loses its elasticity.

Synoptic link

You can read more about arteries in Key Area 2.5 (page 53).

Thrombosis

- Atheromas may rupture, damaging the endothelium of the artery.
- This damage releases clotting factors that trigger a series of reactions resulting in the conversion of the inactive enzyme **prothrombin** to its active form **thrombin**.
- Thrombin then causes molecules of the soluble plasma protein fibrinogen to form threads of insoluble fibrin protein.
- The fibrin threads form a meshwork that traps blood cells and clots the blood, seals the wound and provides a scaffold for the formation of scar tissue.
- The formation of a clot (thrombus) is referred to as **thrombosis**.
- In some cases a thrombus may break loose from the blood vessel wall, forming an **embolus** which travels through the bloodstream and could block a smaller blood vessel further on in the circulatory system.
- If thrombosis occurs in a coronary artery, it may lead to a **myocardial infarction (MI)**, commonly known as a heart attack, as shown in Figure 2.28.
- A thrombosis in an artery in the brain may lead to a stroke.
- Thrombosis normally results in the death of some of the tissue served by the blocked artery as the cells are deprived of oxygen.

Key terms

Fibrinogen Blood protein that is converted to fibrin during the blood-clotting process.

Fibrin Threads of protein that help to form a blood clot.

Thrombosis Formation of a blood clot within a blood vessel.

Embolus A detached mass of material, such as a thrombus, carried by blood circulation.

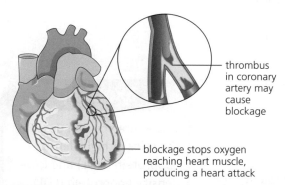

thrombus in coronary artery may cause blockage

blockage stops oxygen reaching heart muscle, producing a heart attack

Figure 2.28 Myocardial infarction (MI); a thrombus from somewhere in the circulatory system becomes trapped in a coronary artery, causing a blockage and depriving the heart muscle cells of oxygen

Peripheral vascular disorders

- Peripheral vascular disease is the narrowing of arteries, due to atherosclerosis, in arteries other than those of the heart or brain.
- The arteries to the legs are most commonly affected.
- A **deep vein thrombosis (DVT)** is the formation of a thrombus (blood clot) that forms in a deep vein, most commonly in the lower leg, as shown in Figure 2.29.
- In deep vein thrombosis, pain is experienced in the leg muscles due to a limited supply of oxygen.

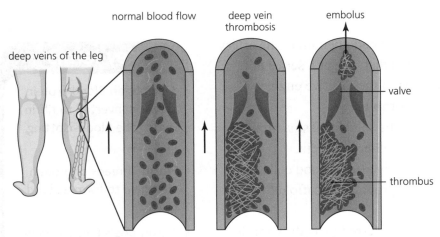

normal blood flow

deep vein thrombosis

embolus

deep veins of the leg

valve

thrombus

Figure 2.29 Formation of a deep vein thrombosis (DVT) in the lower leg

■ A **pulmonary embolism** in the lungs occurs when part of a thrombus breaks free and travels through the bloodstream to the pulmonary artery where it can cause a blockage, resulting in chest pain and breathing difficulties.

Control of cholesterol levels

■ Cholesterol is a type of lipid found in the cell membrane.
■ It is also used to make the sex hormones testosterone, oestrogen and progesterone.
■ Cholesterol is synthesised by all cells, although 25% of total production takes place in the liver.
■ A diet that is high in saturated fats or cholesterol causes an increase in cholesterol levels in the blood.

Transport of cholesterol

■ **Lipoproteins** contain lipids and proteins and are responsible for the transport of cholesterol in the blood.
■ **High-density lipoprotein (HDL)** transports excess cholesterol from the body cells to the liver for elimination, preventing the accumulation of cholesterol in the blood.
■ **Low-density lipoprotein (LDL)** transports excess cholesterol to body cells.
■ Most cells have **LDL receptors** that take LDL into the cell, where it releases the cholesterol.
■ Once a cell has sufficient cholesterol, a negative feedback system inhibits the synthesis of new LDL receptors.
■ LDL circulates in the blood where it may deposit the cholesterol in the arteries, forming atheromas, as shown in Figure 2.30.

> **Exam tip**
>
> In your exam you may be asked to give examples of peripheral vascular disorders: deep vein thrombosis (DVT) and pulmonary embolism.

> **Synoptic link**
>
> You can read more about veins in Key Area 2.5 (page 53).

> **Exam tip**
>
> Make sure you can describe the roles of high-density lipoprotein (HDL) and low-density lipoprotein (LDL).

> **Key terms**
>
> Lipoprotein An assembly of protein with lipid that enables movement of lipids in water and through membranes.
>
> LDL receptors Protein receptors that recognise LDLs and encourage their uptake.

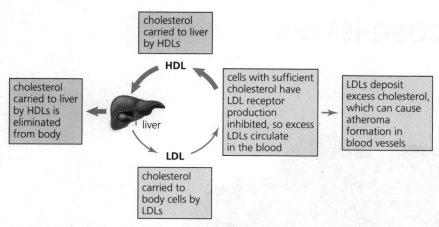

Figure 2.30 Roles of HDL and LDL in the transport of cholesterol

Reducing cholesterol levels

- A higher ratio of HDL to LDL results in lower blood cholesterol levels and a reduced chance of atherosclerosis.
- Regular physical activity tends to raise HDL levels and so lowers blood cholesterol.
- Dietary changes aim to reduce levels of total fat in the diet and to replace saturated with unsaturated fats.
- Drugs such as **statins** reduce blood cholesterol by inhibiting the synthesis of cholesterol by liver cells.

Do you know?

1 Write notes on the process of atherosclerosis and its effect on arteries and blood pressure. [5]
2 Give an account of the rupture of an atheroma, leading to the formation of a thrombus. [5]

2.8 Blood glucose levels and obesity

You need to know

- that chronic elevated blood glucose levels lead to atherosclerosis and blood vessel damage
- about pancreatic receptors and the role of hormones in negative feedback control of blood glucose through insulin, glucagon and adrenaline
- the features of type 1 and type 2 diabetes
- the measurement of obesity and its impact on health

Blood glucose levels and vascular disease

- Chronic elevated blood glucose levels may lead to blood vessel damage:
  - ☐ Untreated diabetes causes chronic elevation of blood glucose levels.
  - ☐ This means that endothelial cells lining blood vessels absorb more glucose than normal, which causes damage.
- Chronic elevated blood glucose levels may also lead to atherosclerosis:
  - ☐ Atherosclerosis may lead to cardiovascular disease (CVD), stroke or peripheral vascular disease.
  - ☐ Peripheral vascular disease affects blood vessels leading to the arms, hands, legs, feet and toes.
  - ☐ Damage to small blood vessels by elevated glucose levels may result in **haemorrhaging** of blood vessels in the **retina**, **renal failure** or **peripheral nerve dysfunction**.

Key terms

Haemorrhage An escape of blood from a ruptured blood vessel.

Retina The light-sensitive layer of the back of the eye.

Renal failure Kidney failure.

Peripheral nerve dysfunction The result of damage to your peripheral nerves, which may impair sensation, movement or gland or organ function.

Exam tip

In your exam you may be asked to describe the effects of chronic elevated blood glucose levels on the circulatory system. These are atherosclerosis, CVD, stroke, peripheral vascular disease, blood vessel damage and hypertension.

Synoptic link

You can read more about CVD, strokes and peripheral vascular disease in Key Area 2.7 (page 62).

Regulation of blood glucose levels

- Blood glucose concentration is monitored by receptors in the pancreas and maintained within tolerable limits by **negative feedback control** involving the hormones insulin, glucagon and adrenaline.
- The pancreas controls blood glucose by manufacturing insulin and glucagon, which act antagonistically. The hormones are transported in the blood to the liver.
- Pancreatic receptors respond to raised blood glucose levels by increasing the secretion of insulin from the pancreas.
  - □ The insulin makes the liver cells more permeable to glucose and activates the conversion of glucose to glycogen to be stored in the liver cells.
  - □ This decreases blood glucose concentration.
- Pancreatic receptors respond to lowered blood glucose levels by increasing the secretion of glucagon from the pancreas.
  - □ Glucagon activates the conversion of glycogen to glucose in the liver.
  - □ The glucose is released, increasing blood glucose concentration.
- Figure 2.31 provides a summary of the negative feedback regulation of blood glucose levels.

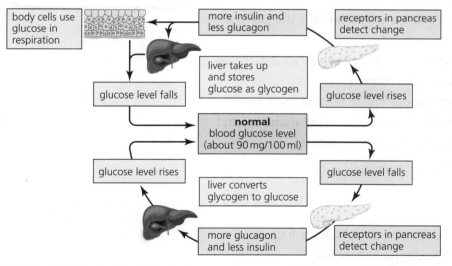

Figure 2.31 Control of blood glucose concentration by negative feedback

- During exercise and in fight-or-flight responses, glucose concentration in the blood is raised by the hormone adrenaline.
- Adrenaline is released from the adrenal glands located on the top of each kidney.
- Adrenaline stimulates glucagon secretion and inhibits insulin secretion.

Type 1 and type 2 diabetes

- Vascular disease can be a chronic complication of diabetes.
- **Type 1 diabetes** usually shows up in childhood:
  - ☐ Affected individuals cannot produce insulin.
  - ☐ Treatment involves regular injections with carefully measured doses of insulin.
- **Type 2 diabetes** typically develops later in an individual's life:
  - ☐ Being overweight increases the chances of developing type 2 diabetes.
  - ☐ Affected individuals produce insulin but their liver cells are insulin-resistant. This means that they have a smaller number of insulin receptors on their surface, which leads to failure to take up glucose and convert it to glycogen.
- In both types of diabetes:
  - ☐ Blood glucose concentration rises rapidly after a meal.
  - ☐ The kidneys are unable to reabsorb all the glucose, resulting in glucose being excreted in the urine.
  - ☐ Testing urine for glucose is often used as an indicator of diabetes.
- The **glucose tolerance test** is used to diagnose diabetes:
  - ☐ The blood glucose concentrations of the individual are first measured after fasting.
  - ☐ The individual then drinks a standard dose of glucose solution.
  - ☐ Changes in their blood glucose concentration are measured for at least the next 2 hours to determine how quickly the glucose is cleared from the blood.
  - ☐ The blood glucose concentration of an individual with diabetes usually starts at a higher level than normal.
  - ☐ During the test, their blood glucose concentration increases to a much higher level than that of someone without diabetes, and takes longer to return to its starting concentration.

Obesity

- **Obesity** may impair health and is a major risk factor for cardiovascular disease (CVD) and type 2 diabetes.
- It is characterised by excess body fat in relation to lean body tissue, such as muscle.
- **Body mass index (BMI)** is commonly used to measure obesity.
- BMI is a measurement of body fat based on height and weight.
- BMI = body mass (kg) divided by height (m) squared.

Exam tip

In your exam you may be asked to describe the difference between type 1 and type 2 diabetes. Individuals with type 1 diabetes are unable to produce insulin, while individuals with type 2 diabetes produce insulin but their cells are less sensitive to it.

Key term

Glucose tolerance test Blood test that measures the body's ability to maintain a normal blood glucose level.

- BMI can be used to categorise individuals as obese, overweight, normal or underweight, as shown in Table 2.1. A BMI greater than $30\,\text{kg}\,\text{m}^{-2}$ is used to indicate obesity.
- One disadvantage of the BMI measurement is that someone may be wrongly classified as overweight or obese when additional mass is actually muscle or bone and not fat.

Technique

Measuring BMI is a technique you are expected to be familiar with for your exam. Make sure you are familiar with the formula for BMI (BMI = body mass (kg) divided by height (m) squared) and be aware of the problems that arise when using the index with individuals who possess extremely muscular physiques.

Table 2.1 **BMI measurements used to categorise individuals**

BMI range	Category
<18.5	Underweight
18.5–24.9	Normal
25–29.9	Overweight
30+	Obese

- Obesity is linked to high-fat diets and a decrease in physical activity.
- In preventing and tackling obesity, the energy intake in the diet should be reduced by limiting fats and free sugars.
- Fats have a high calorific value per gram and free sugars require no metabolic energy to be expended in their digestion.
- Exercise can be used to increase energy expenditure and preserve lean tissue.
- Exercise can also help to reduce risk factors for cardiovascular disease (CVD) by keeping weight under control, minimising stress, reducing hypertension and improving HDL blood lipid profiles.

Synoptic link

You can read more about HDL to LDL ratios in Key Area 2.7 (page 62).

Do you know?

1 Write notes on the control of blood glucose following an increase in blood glucose concentration after a meal. [4]
2 Write notes on the control of blood glucose following a decrease in blood glucose concentration after exercise. [4]

Area 2 assessment
Physiology and health

Answer on separate sheets of paper. Mark your own work at: hoddereducation.co.uk/needtoknow/answers

1 The testes are male reproductive organs and they contain germline cells.

   a Describe the location and function of the germline cells in the testes.   [1]

   b Name the cells in the testes that secrete testosterone.   [1]

   c Give *one* function of testosterone.   [1]

   d Describe the roles of the secretions from the seminal vesicles and prostate gland.   [1]

2 Hormones control the onset of puberty, sperm production and the menstrual cycle.

   a Describe the role of the following hormones in male reproduction:

      i   FSH   [1]

      ii ICSH   [1]

   b Describe the role of the following hormones in the menstrual cycle:

      i   FSH   [1]

      ii LH   [1]

   c Describe the role of progesterone in the menstrual cycle.   [1]

   d Explain what prevents further development of follicles when an embryo is developing
      in the uterus.   [1]

3 Infertility treatments and contraception are based on the biology of fertility.

   a Give the meaning of the terms 'cyclical fertility' and 'continuous fertility'.   [1]

   b Describe the process of artificial insemination in the treatment of infertility.   [1]

   c Describe the steps involved in the ICSI technique.   [1]

   d Explain why pre-implantation genetic diagnosis (PGD) is used in conjunction with IVF.   [1]

   e Explain how the contraceptive pill, based on synthetic hormones, works.   [1]

4 Antenatal screening uses tests to identify the risk of the presence of a disorder before birth and
   postnatal screening involves diagnostic testing of newborn babies.

   a Give *one* example of antenatal screening procedures.   [1]

   b Give *one* use of ultrasound imaging obtained by ultrasound scanners.   [1]

   c Explain the difference between a screening test and a diagnostic test.   [1]

   d Give *one* advantage of chorionic villus sampling over amniocentesis as a diagnostic technique. [1]

   e Give *one* characteristic of a family tree that would enable a geneticist to establish whether a
      condition or disorder was autosomal recessive.   [1]

   f Give *one* characteristic of a family tree that would enable a geneticist to establish whether a
      condition or disorder was sex-linked recessive.   [1]

5 Blood vessels are tubes with walls composed of different tissues, depending on the function of
   the vessel.

   a Describe how the presence of muscle tissue in the artery wall helps to control the flow
   of blood around the body. [1]

   b Describe the role of the elastic fibres in the wall of an artery. [1]

   c Give *one* structural difference between an artery and a vein. [1]

   d Describe the role of tissue fluid. [1]

   e Describe the role of lymph vessels. [1]

6 The cardiac cycle is the pattern of contraction and relaxation of the heart muscle in one complete
   heartbeat.

   a Give the meaning of the term 'cardiac output'. [1]

   b Describe the role of the atrioventricular node (AVN). [2]

   c Describe the action of the sympathetic nervous system in the control of the cardiac cycle. [1]

   d Describe the state of the heart valves during ventricular systole. [2]

7 Atherosclerosis is the root cause of various cardiovascular diseases.

   a Describe the process of atherosclerosis. [1]

   b Describe the effect of an atheroma. [2]

   c Name *one* cardiovascular disease associated with atherosclerosis. [1]

   d Describe the events leading to the formation of a thrombus, following the release of clotting
   factors after the rupture of an atheroma. [2]

   e Describe how thrombosis can lead to myocardial infarction (MI). [1]

   f Explain the benefit of a higher ratio of HDL to LDL in blood plasma. [2]

8 The maintenance of blood glucose concentration within tolerable limits is an example of
   homeostasis.

   a Name *one* condition associated with chronic elevated blood glucose levels. [1]

   b Give *one* damaging effect caused by the haemorrhage of
   small blood vessels due to elevated glucose levels. [1]

   c Name the type of control that maintains blood glucose levels within tolerable limits. [1]

   d Describe the role of the hormone insulin. [1]

   e Describe the role of the hormone glucagon. [1]

   f Explain the role of the hormone adrenaline in the control of blood glucose concentration. [1]

   g Describe the difference between type 1 and type 2 diabetes. [1]

9 Obesity may impair health and is a major risk factor for cardiovascular disease and type 2 diabetes.

   a Give the equation that is used to calculate BMI. [1]

   b State *one* way in which exercise reduces the risk factors for cardiovascular disease (CVD). [1]

Extended response

10 Discuss procedures that can be used to treat infertility. [10]

3 Neurobiology and immunology

3.1 Divisions of the nervous system and neural pathways

You need to know

- the structure of the central nervous system (CNS) and the peripheral nervous system (PNS)
- the structure and function of converging, diverging and reverberating neural pathways

Structure of the nervous system

- The nervous system can be divided into the **central nervous system (CNS)** and the **peripheral nervous system (PNS)**, as shown in Figure 3.1. The functions of the different parts of the nervous system are shown in Figure 3.2.
- The **somatic nervous system (SNS)** controls voluntary actions by skeletal muscles and contains sensory and motor **neurons**:
  - □ **Sensory neurons** are nerve cells that carry electrical impulses from sense organs to the CNS.
  - □ **Motor neurons** are nerve cells that carry electrical impulses from the CNS to muscles and glands.
- The **autonomic nervous system (ANS)** controls involuntary actions by glands, smooth muscle and cardiac muscle.
- The ANS consists of the **sympathetic** and **parasympathetic nervous systems**, which are **antagonistic** to each other:
  - □ The sympathetic nervous system speeds up heart and breathing rate while slowing down **peristalsis** and the production of intestinal secretions.
  - □ The parasympathetic nervous system slows the heart and breathing rate but speeds up digestive processes such as intestinal secretions and peristalsis.

Exam tip

In your exam you may be asked to describe the difference between somatic and autonomic actions. Somatic actions are voluntary and brought about by skeletal muscle; autonomic actions are involuntary and brought about by smooth muscle, cardiac muscle or glands.

Key terms

Neuron A conducting nerve cell.

Antagonistic Having opposing effects.

Peristalsis Involuntary wave-like muscular contractions that push food through the digestive system.

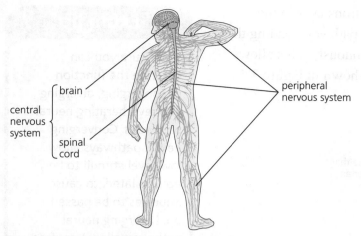

Figure 3.1 Parts of the nervous system

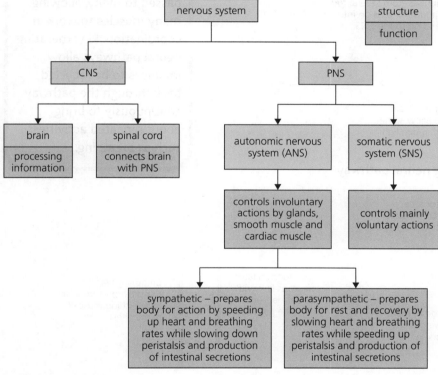

Figure 3.2 Summary of the structure and function of the nervous system

Synoptic links

You can read more about the structure and function of the brain in Key Area 3.2 (page 75), of neurons in Key Area 3.4 (page 79), and about the autonomic nervous system (ANS) and the cardiac conducting system in Key Area 2.6 (page 57).

Converging, diverging and reverberating neural pathways

■ Neurons that are connected to each other through **synapses** form **neural pathways** through the nervous system.

☐ In a **converging neural pathway**, impulses from several neurons travel to a single neuron. This increases their sensitivity to excitatory or inhibitory signals, as shown in Figure 3.3.

☐ In a **diverging neural pathway**, impulses from one neuron travel to several neurons, affecting more than one destination at the same time, as shown in Figure 3.4.

Key term

Synapses Tiny gaps between neurons.

☐ In a **reverberating neural pathway**, neurons later in the pathway link with neurons earlier in the pathway, sending the impulse back through the pathway continuously. This allows repeated stimulation of the pathway, as shown in Figure 3.5.

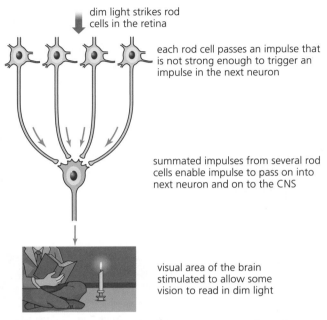

Figure 3.3 **Example of a converging neural pathway**

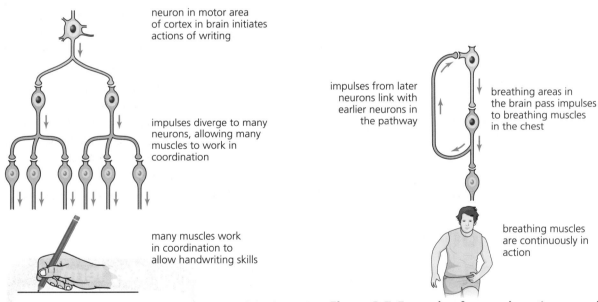

Figure 3.4 **Example of a diverging neural pathway**   Figure 3.5 **Example of a reverberating neural pathway**

Do you know?

1 Describe the structure and function of the autonomic nervous system (ANS). [6]

2 Give an account of the structure and function of neural pathways. [6]

3.2 Cerebral cortex

You need to know
- the function of the cerebral cortex
- the function of the corpus callosum

The brain

- The brain is the complex central processing centre of the nervous system (see Figure 3.6).
- The **cerebrum** is the largest and uppermost portion of the brain.
- The cerebrum consists of the right and left **cerebral hemispheres** and accounts for two-thirds of the total weight of the brain.

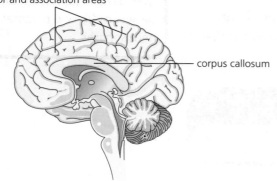

cerebral cortex containing sensory, motor and association areas

corpus callosum

Figure 3.6 Vertical section of the human brain

Function of the cerebral cortex

- The **cerebral cortex** is the thin outer layer of the cerebrum.
- It is the centre of conscious thought.
- It is also important in the recall of memories, and alters behaviour in the light of experience.
- Brain functions in the cerebral cortex are localised, which means that certain parts of the cortex have a specific function.
- The cerebral cortex contains **sensory**, **motor** and **association areas**:
  - Sensory areas receive nervous impulses from the sense organs.
  - Motor areas send nervous impulses to the muscles and glands.
  - Association areas are involved in language processing, personality, imagination and intelligence, as shown in Figure 3.7.

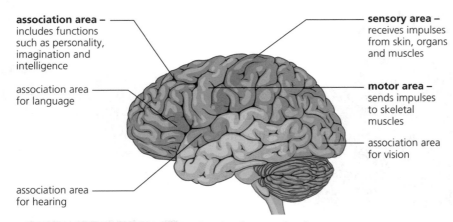

association area – includes functions such as personality, imagination and intelligence

association area for language

association area for hearing

sensory area – receives impulses from skin, organs and muscles

motor area – sends impulses to skeletal muscles

association area for vision

Figure 3.7 Localisation of function in the cerebral cortex

Function of the corpus callosum

- Information from one side of the body is processed in the opposite side of the cerebrum.
- The left cerebral hemisphere deals with information from the right visual field and controls the right side of the body.
- The right cerebral hemisphere deals with information from the left visual field and controls the left side of the body.
- The transfer of information between the cerebral hemispheres occurs through the band of tissue that connects them, called the **corpus callosum**.

Key term

Corpus callosum A bridge of tissue, deep in the brain, that transfers information between the two cerebral hemispheres.

Do you know?

1   Give an account of the role of the cerebral cortex.   [6]

3.3 Memory

You need to know

- that memory involves encoding, storage and retrieval of information
- that the sensory memory retains all the visual and auditory input received for a few seconds
- the features of short-term and long-term memory

Memory

- Memories include past experiences, knowledge and thoughts.
- Memory is the ability to recover information about past events or knowledge.
- **Short-term memory (STM)** recovers memories of recent events, while **long-term memory (LTM)** is concerned with recalling the more distant past.
- Memory involves **encoding**, **storage** and **retrieval** of information:
  - ☐ Encoding information involves converting information into a form that can be received and stored by the brain.
  - ☐ Storage involves holding the information within the memory.
  - ☐ Retrieval is the ability to recall information from the memory when required.

Sensory memory

- All information entering the brain passes through **sensory memory**.
- Sensory memory retains all the visual and auditory input received for a few seconds, but only selected images and sounds are then encoded into STM.

Short-term memory

- Memory span is the number of discrete items, such as letters, words or numbers, that the STM can hold.
- STM has a limited capacity and can hold about seven to nine items for no more than 20 or 30 seconds at a time.
- The capacity of STM can be improved by 'chunking', which involves grouping items together to make a single item.
- Items can be retained in the STM by **rehearsal** or lost by **displacement** or **decay**.
- The **serial position effect** is the tendency of a person to recall those items in a group that come first (primacy) and last (recency) best and to forget the intermediate (middle) items.
- STM can also process data, to a limited extent, as well as store it. This is called the '**working memory model**':
  - ☐ 'Working memory' is the thinking skill that focuses on memory in action.
  - ☐ This is the ability to remember and use relevant information while in the middle of an activity.
  - ☐ For example, an individual is using their working memory as they recall the steps of a recipe while cooking a meal.
- Figure 3.8 shows the relationships between the sensory memory, the STM and the LTM.

Key terms

Encoding Involves converting nerve signals into a form that can be received and interpreted by the brain.

Storage The holding of information within the memory.

Retrieval The recall of information from memory.

Sensory memory Storage of sensory input which lasts a few seconds and retains all of the visual or auditory inputs.

Key terms

Rehearsal Practice by repetition of an item of information.

Displacement The loss of items from STM.

Decay Degradation of information in STM.

Working memory model A system for temporarily storing and managing the information required to carry out complex cognitive tasks, such as learning, reasoning and comprehension.

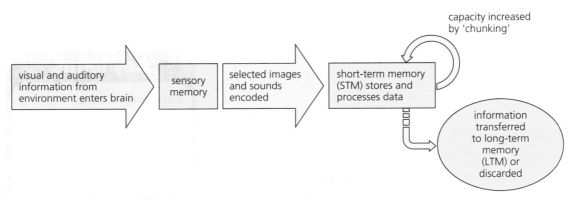

Figure 3.8 Summary of short-term memory (STM)

Long-term memory

- Long-term memory (LTM) has an unlimited capacity and holds information for a long time.
- Information can be transferred from STM to LTM by rehearsal, **organisation** and **elaboration**:
  - ☐ Organisation involves placing the information into an organised framework such as a table or chart.
  - ☐ Elaboration involves building more detail around the core information.
- Information can be encoded by shallow processing or deep, elaborative processing.
  - ☐ Rehearsal is regarded as a shallow form of encoding information into LTM.
  - ☐ Elaboration is regarded as a deeper form of encoding which leads to improved information retention.
- Retrieval from LTM is aided by the use of contextual cues. Contextual cues relate to the time and place when the information was initially encoded into LTM.

> ### Key terms
>
> **Organisation** Method of encouraging transfer of information to LTM by adding structure.
>
> **Elaboration** Method of encouraging transfer of information to LTM by using detail to add meaning.

> ### Exam tip
>
> Ensure you can name the methods of transferring information from the STM to the LTM.

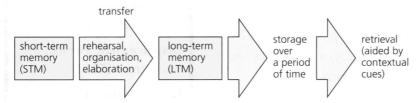

Figure 3.9 Summary of transfer, storage and retrieval in long-term memory (LTM)

> ## Do you know?
>
> 1 Give an account of short-term memory. [6]
> 2 Give an account of long-term memory. [5]
> 3 Give an account of the transfer of information from short-term to long-term memory. [6]

3.4 Cells of the nervous system and neurotransmitters at synapses

You need to know

- the structure and function of neurons
- about the action of neurotransmitters at synapses
- the effects of neurotransmitters on mood and behaviour
- examples of neurotransmitter-related disorders and their treatments
- the mode of action of recreational drugs

Neurons

- **Neurons** are nerve cells adapted to carry electrical impulses.
- They have long fibres to carry these impulses through the nervous system.
- The two types of neuron are sensory and motor neurons, as shown in Figure 3.10.
- Neurons consist of a cell body and fibres called **dendrites** and **axons**:
  - ☐ Dendrons and dendrites are fibres that conduct impulses towards the cell body of the neuron.
  - ☐ Axons conduct impulses away from the cell body, and usually branch at their ends.
  - ☐ Axons are surrounded by a myelin sheath (as shown in Figure 3.10), which insulates the axon and increases the speed of impulse conduction.

Key terms

Dendrites Neural fibres that conduct impulses towards the cell body.

Axons Neural fibres that conduct impulses away from the cell body.

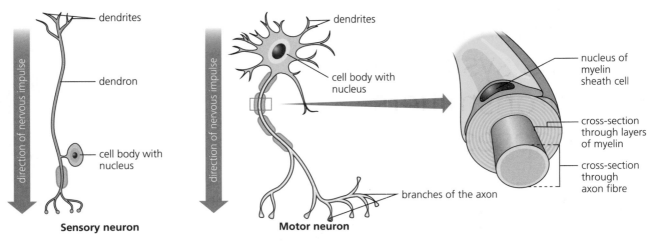

Figure 3.10 Types of neuron, showing the position and detail of the myelin sheath

Myelination and glial cells

- **Myelination** continues from birth to adolescence.
- Responses to stimuli in the first 2 years of life are not as rapid or coordinated as those of an older child or adult due to myelination being incomplete.
- Certain diseases such as multiple sclerosis destroy the myelin sheath, causing a loss of coordination.
- **Glial cells** are associated with neurons. They produce the myelin sheath and provide physical support for neurons, as shown in Figure 3.11.

> **Key term**
>
> **Myelination** Insulation of nerve fibres with myelin sheaths.

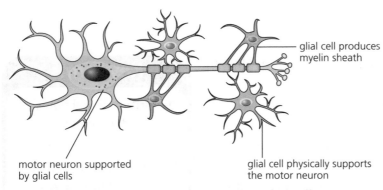

Figure 3.11 Motor neuron with supporting glial cells

Action of neurotransmitters at synapses

- Synapses are tiny gaps between neurons.
- Neurons connect with other neurons or muscle fibres at a **synaptic cleft**.
- Neurotransmitters relay impulses across the synaptic cleft.

> **Key term**
>
> **Synaptic cleft** Gap between neurons at a synapse.

- Neurotransmitters are stored in **vesicles** in the axon endings of the **presynaptic neuron**.
- Neurotransmitters are released into the synaptic cleft on arrival of an electrical impulse, then diffuse across the synaptic cleft and bind to receptor proteins on the membrane of the **postsynaptic neuron**.
- Receptors in the postsynaptic membrane recognise the neurotransmitter molecules, which triggers a nervous impulse that passes along the dendrite of the postsynaptic neuron, as shown in Figure 3.12.

Key terms

Vesicle (neurotransmitter)
Tiny vacuole in presynaptic neurons containing neurotransmitter.

Presynaptic neuron
Neuron that releases neurotransmitters into the synaptic cleft on arrival of an electrical impulse.

Postsynaptic neuron
Neuron that contains membrane receptors for neurotransmitters.

Exam tip

Make sure you can describe how neurotransmitters are involved in the passage of nervous impulses across the synapse. Neurotransmitters released from vesicles diffuse across the synaptic cleft and bind to membrane receptors on the postsynaptic neuron.

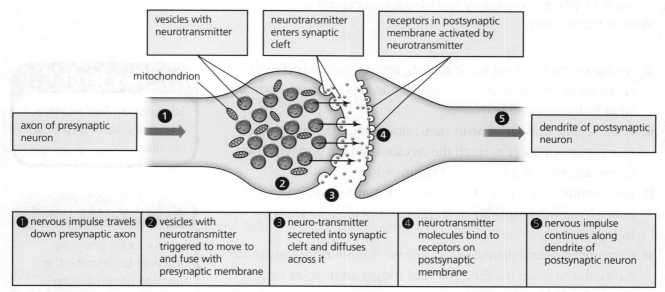

Figure 3.12 Stages of nervous transmission at a synapse

- The neurotransmitters need to be removed quickly by enzymes or reabsorption in order to prevent continuous stimulation of postsynaptic neurons.
- This allows the system to respond to new signals and makes precise control possible.

Exam tip

Remember that neurotransmitters can be removed through degradation by enzymes or by reabsorption by the presynaptic membrane.

Types of signal and the transmission of impulses

- The type of receptor on the postsynaptic membrane determines whether the signal is excitatory or inhibitory, as shown in Figure 3.13.
- It is the sum of the excitatory and inhibitory signals that determines the overall effect.

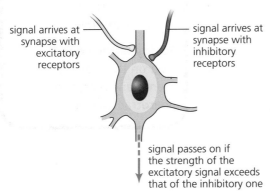

Figure 3.13 Effect of excitatory and inhibitory receptors on impulse transmission

- Synapses can filter out weak stimuli, arising from insufficient secretion of neurotransmitters, which prevents very weak stimuli from bringing about responses.
- A minimum number of neurotransmitter molecules must attach to receptors in order to reach the threshold on the postsynaptic membrane required to transmit the impulse.
- **Summation** of a series of weak stimuli can release enough neurotransmitter to trigger an impulse, as shown in Figure 3.14.
- Convergent neural pathways can release enough neurotransmitter molecules to reach the threshold and trigger an impulse (see Figure 3.3 in Key Area 3.1).

Key term

Summation The additive effect of several electrical impulses.

Exam tip

In your exam you may be asked to describe the function of converging neural pathways. These pathways allow low-level stimuli to be summated, to cause impulses to be passed on.

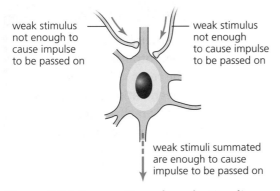

Figure 3.14 Summation of weak stimuli

Effect of neurotransmitters on mood and behaviour

- **Endorphins** are neurotransmitters that stimulate neurons involved in reducing the intensity of pain.
- The production of endorphins increases in response to severe injury, prolonged and continuous exercise, stress and certain foods.
- Increased levels of endorphins are linked to the feelings of pleasure obtained from activities such as eating, sex and prolonged exercise.
- **Dopamine** is a neurotransmitter that induces feelings of pleasure and reinforces particular behaviours by activating the reward pathway in the brain.
- The **reward pathway** involves neurons that secrete or respond to dopamine.
- The reward pathway is activated when an individual engages in a behaviour that is beneficial to them, for example, eating when hungry, and increases the chances of that particular behaviour happening again.

Neurotransmitter-related disorders

- Some drugs affect the way in which neurotransmitters function and can be used to treat neurotransmitter-related disorders, as shown in Figure 3.15.

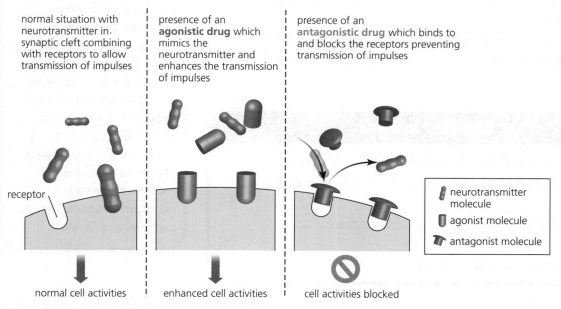

Figure 3.15 Action of agonistic and antagonistic drugs

- **Agonists** are chemicals that bind to and stimulate specific receptors and mimic the action of a neurotransmitter at a synapse, so enhancing their action.

- **Antagonists** are chemicals that bind to specific receptors, blocking the action of a neurotransmitter at a synapse and so preventing nerve impulses passing across synapses.
- Other drugs act by inhibiting the enzymes that degrade the neurotransmitter in the synaptic cleft or by inhibiting the reuptake of the neurotransmitter at the synapse, causing an enhanced effect.

Recreational drugs

- Many recreational drugs affect neurotransmission by acting as agonists or antagonists in the reward pathway of the brain.
- Recreational drugs affect neurotransmission at synapses in the brain, altering an individual's mood, **cognition**, **perception** and behaviour.
- **Drug addiction** is caused by repeated use of drugs that act as antagonists:
  - ☐ Antagonist drugs block specific receptors, causing the nervous system to increase both the number and sensitivity of these receptors.
  - ☐ This sensitisation leads to addiction, where the individual craves more of the drug.
- **Drug tolerance** is caused by repeated use of drugs that act as agonists:
  - ☐ Agonist drugs stimulate specific receptors, causing the nervous system to decrease both the number and sensitivity of these receptors.
  - ☐ This desensitisation leads to drug tolerance, where the individual must take more of the drug to produce the desired effect.

Key terms

Cognition Process of acquiring knowledge and understanding through thought, experience, and the senses.

Perception Capacity to take in and make sense of sensory information.

Exam tip

Make sure you can explain the difference between drug addiction and drug tolerance in terms of neural receptors. Sensitisation (increase in neural receptors) through exposure to antagonist chemicals can lead to addiction. Desensitisation (decrease in neural receptors) through exposure to agonist drugs can lead to tolerance.

Do you know?

1 Give an account of the structure and function of neurons. [5]
2 Give an account of the transmission of a nervous impulse at a synapse. [4]
3 Give an account of the role of neurotransmitters at a synapse. [6]
4 Give an account of agonist drugs. [4]

3.5 Non-specific body defences

You need to know

- examples of physical and chemical defences of the body
- about the inflammatory response
- the role and action of phagocytes

Physical and chemical defences

- A **pathogen** is a bacterium, virus or other organism that can cause disease.
- **Non-specific defences** are bodily mechanisms that do not target specific pathogens and do not give long-lasting protection or immunity.
- Non-specific defences can be physical and chemical:
  - ☐ Closely-packed **epithelial cells** are found in the skin and inner linings of the digestive and respiratory systems, and form a physical barrier against the entry of pathogens, as shown in Figures 3.16 and 3.17.
  - ☐ Chemical secretions such as tears, saliva, mucus and stomach acid are produced against invading pathogens.

> ### Key terms
>
> **Non-specific defence** A general response to infection that does not target a particular pathogen.
>
> **Epithelial cells** Cells that line tubes and surfaces in the body.

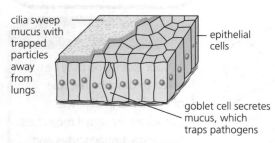

cilia sweep mucus with trapped particles away from lungs

epithelial cells

goblet cell secretes mucus, which traps pathogens

Figure 3.16 Ciliated epithelium from the respiratory system

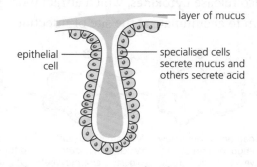

layer of mucus

epithelial cell

specialised cells secrete mucus and others secrete acid

Figure 3.17 Gastric gland in the stomach epithelium

The inflammatory response

> ### Key term
>
> **Mast cell** Type of white blood cell that produces histamine in response to tissue damage.

- The inflammatory response is a defence mechanism triggered by damage to living tissue.
- **Histamine** is released by **mast cells**, causing vasodilation and increased permeability of capillaries.

■ The increased blood flow leads to an accumulation of **phagocytes** and clotting elements at the site of infection, as shown in Figure 3.18.

events of damage, inflammation and repair processes occurring through time →

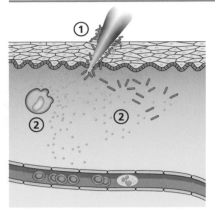 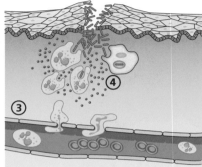 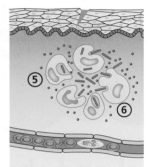

❶ needle carrying bacteria breaks skin

❷ mast cells release histamine

❸ histamine causes blood capillary to vasodilate and become more permeable

❹ increased permeability of capillary allows phagocytes and clotting elements to accumulate near the infection site

❺ accumulated phagocytes engulf and destroy bacteria by phagocytosis, and clotting elements promote wound healing

❻ phagocytes release cytokines, which act as signal molecules to attract more phagocytes to the site of infection

Figure 3.18 The inflammatory response at a site of infection

Action of phagocytes

■ Phagocytes recognise pathogens and destroy them by **phagocytosis**.

■ Phagocytosis involves engulfing pathogens into a vesicle and destroying them with powerful digestive enzymes contained in **lysosomes**, as shown in Figure 3.19.

■ Phagocytes also release **cytokines**, which attract more phagocytes and cause them to accumulate at the site of infection.

pathogen ❶   ❷   lysosome ❸   ❹

pathogen detected by phagocyte

pathogen completely engulfed into a vesicle inside phagocyte

lysosomes fuse with vesicle and release digestive enzymes that destroy pathogen

digested products diffuse into cytoplasm of phagocyte, providing nutrition

Figure 3.19 Stages in phagocytosis

Do you know?

1 Give an account of the inflammatory response. [4]
2 Give an account of the role of phagocytes. [5]

3.6 Specific cellular defences against pathogens

You need to know
- the role and action of lymphocytes
- about the secondary response

Action of lymphocytes

- **Lymphocytes** are the white blood cells involved in the specific immune response.
- These cells respond to specific **antigens** on invading **pathogens**.
- Antigens are molecules (often proteins) located on the surface of cells which trigger a **specific immune response**.
- During development, stem cells differentiate into many types of lymphocyte.
- Each type of lymphocyte has a single type of membrane receptor which is specific for one antigen.
- When the specific antigen binds to its specific receptor, repeated lymphocyte division occurs, which results in the formation of a **clonal population** of identical lymphocytes, as shown in Figure 3.20.

Key terms

Pathogen A bacterium, virus, or other microorganism that can cause disease.

Clonal population A group of genetically identical cells produced from one parent cell.

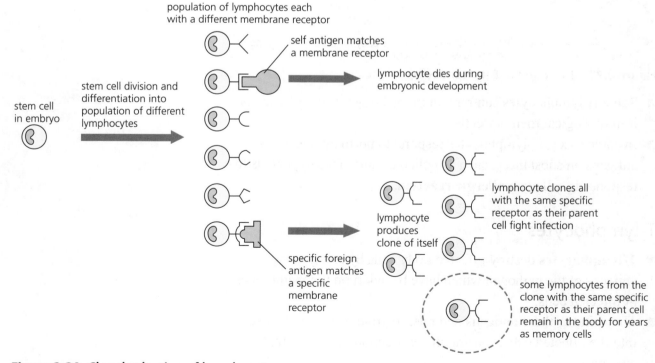

Figure 3.20 Clonal selection of lymphocytes

- Some lymphocytes from the clone can remain in the body for many years as immunological **memory cells**.

Types of lymphocyte

- There are two types of lymphocyte, called **B-lymphocytes** and **T-lymphocytes**.

B-lymphocytes

- **B-lymphocytes** produce **antibodies** against antigens.
- Antibodies are Y-shaped proteins that have receptor-binding sites specific to a particular antigen on a pathogen.
- Antibodies become bound to antigens, inactivating the pathogen.
- The resulting antigen–antibody complex can then be destroyed by phagocytosis, as shown in Figure 3.21.

Key term

Memory cells
Lymphocytes that remember the same pathogen for faster antibody production in future infections.

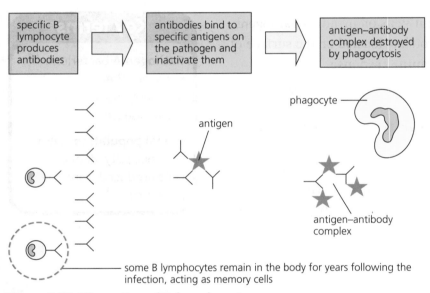

Synoptic link

You can read more about phagocytosis in Key Area 3.5 (page 85).

some B lymphocytes remain in the body for years following the infection, acting as memory cells

Figure 3.21 The action of B-lymphocytes

- Some B-lymphocytes can remain in the body for many years as immunological memory cells.
- In some cases, B-lymphocytes respond to normally harmless antigens on substances, such as pollen or nuts. This hypersensitive response is called an **allergic reaction**.

T-lymphocytes

- **T-lymphocytes** destroy infected body cells by recognising antigens of the pathogen which have been left on the membranes of infected cells.
- They bind to these antigens and release proteins which diffuse into the infected cells, causing the production of self-destructive enzymes.

- This causes programmed cell death, called **apoptosis**.
- The remains of the infected cells are then removed by phagocytosis, as shown in Figure 3.22.

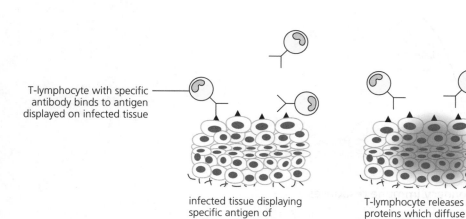
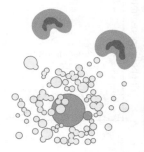

T-lymphocyte with specific antibody binds to antigen displayed on infected tissue

phagocytes remove remains of cells by phagocytosis

infected tissue displaying specific antigen of pathogen involved

T-lymphocyte releases proteins which diffuse into infected tissues

cells release self-destructive enzymes which cause cell death (apoptosis)

Figure 3.22 The action of apoptosis-inducing T-lymphocytes

- T-lymphocytes can normally distinguish between **self-antigens** on the body's own cells and non-self-antigens on infected cells.
- Failure of the regulation of the immune system can lead to T-lymphocytes responding to self-antigens which are found naturally on an individual's body cells.
- The T-lymphocytes then attack the body's own cells, causing **autoimmune diseases** such as type 1 diabetes and **rheumatoid arthritis**.

The secondary response

- Following a primary exposure to an antigen, an individual may become ill and then recover because of their immune response.
- Some of the cloned B- and T-lymphocytes that they produced survive long-term as memory cells.
- When a secondary exposure to the same antigen occurs, these memory cells respond rapidly and give rise to new clones of specific lymphocytes.
- During the **secondary response**, antibody production is greater and more rapid than during the **primary response**, as shown in Figure 3.23.
- Invading pathogens are destroyed before the individual shows any symptoms of the disease.

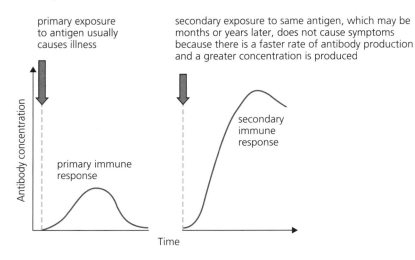

Figure 3.23 The primary and secondary immune responses

- The **human immunodeficiency virus (HIV)** attacks and destroys T-lymphocytes.
- HIV causes depletion of T-lymphocytes which can lead to the development of acquired immune deficiency syndrome (AIDS).
- Individuals with AIDS have a weakened immune system and so are more vulnerable to **opportunistic infections**.

Key term

Opportunistic infections Infections that occur more frequently and are more severe in people with weakened immune systems.

Exam tip

Be clear on the reason why individuals with AIDS are more vulnerable to opportunistic infections. HIV attacks T-lymphocytes, depleting their number and lowering the body's immune-response capabilities.

Do you know?

1 Give an account of the action of B-lymphocytes. [4]
2 Give an account of the action of T-lymphocytes. [5]
3 Give an account of the allergic and autoimmune responses. [4]
4 Describe the role of memory cells in a specific immune response. [6]
5 Describe the difference that would be expected in the response of the immune system to a second exposure to the same pathogen, compared with the first exposure. [3]

3.7 Immunisation

You need to know
- about the process of vaccination
- the role of herd immunity
- the implications of antigenic variation

Vaccination

- **Immunity** is the ability to resist a particular infection or toxin by the action of specific antibodies.
- A vaccine is a biological preparation that produces an immune response without the symptoms of the disease.
- Immunity can be developed by vaccination using antigens from infectious pathogens to create memory cells.
- The antigens used in vaccines can be inactivated pathogen toxins, dead pathogens, parts of pathogens and weakened pathogens.
- Antigens are usually mixed with an **adjuvant** when producing the vaccine.
- An adjuvant is a substance that makes the vaccine more effective, thereby enhancing the immune response and resulting in a higher concentration of antibodies being produced.

Key terms

Immunity The ability of the body to recognise, neutralise, or destroy harmful foreign substances in the body.

Vaccination Using an antigen (made harmless) to produce an immune response and memory cells.

Herd immunity The principle by which many vaccinated individuals protect those who are unvaccinated.

Exam tip

Make sure you can describe what is meant by a vaccine and explain the role of adjuvants for your exam.

Synoptic link

You can read more about memory cells in Key Area 3.6 (page 87).

Herd immunity

- **Herd immunity** occurs when a large percentage of a population is immunised against a disease that can be spread from person to person.
- The immune individuals act as barriers to the spread of infection.
- Establishing herd immunity is important in reducing the spread of diseases.
- Non-immune individuals are protected as there is a lower probability that they will come into contact with infected individuals, as shown in Figure 3.24.

Exam tip

In your exam you may be asked to give the meaning of the terms 'herd immunity' and 'herd immunity threshold' – make sure you revise them.

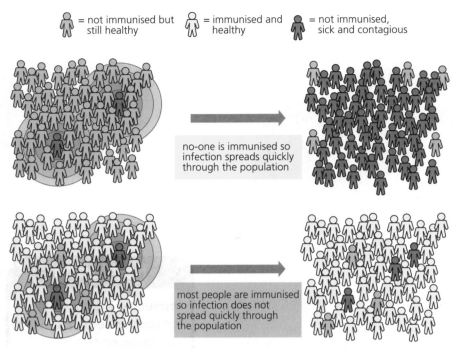

Figure 3.24 The principle of herd immunity is that if enough individuals are immune, they will act as barriers to the spread of disease

- The **herd-immunity threshold** is the percentage of the population that has to be immune to establish herd immunity.
- The threshold depends on the type of disease, the effectiveness of the vaccine and the density of the population.
- Mass vaccination programmes in developed countries are designed to establish herd immunity to a disease.
- Difficulties can arise when vaccines are rejected by a percentage of the population.
- In developing countries, widespread vaccination may not be possible due to poverty, which prevents access to vaccines and structures to supply and administer them.

Antigenic variation

- Some pathogens can change their antigens so that an individual's memory cells are not effective against them.
- This **antigenic variation** occurs in the influenza virus, explaining why it remains a major public health problem and why individuals who are at risk require a vaccination every year.

Key terms

Herd-immunity threshold The percentage of immune people needed for effective herd immunity to occur.

Antigenic variation The ability of some pathogens to change their antigens.

Exam tip

In your exam you may be asked to describe the effect of antigenic variation in pathogens. Remember that when a pathogen changes its antigens, memory cells cannot recognise it and are ineffective against it.

Do you know?

1 Write notes on the production and use of vaccines.   [5]
2 Give an account of the principle of herd immunity.   [5]

3.8 Clinical trials of vaccines and drugs

You need to know
- the role of clinical trials to establish the safety and effectiveness of vaccines and drugs

Clinical trials

- Vaccines and drugs are subjected to **clinical trials** and evaluated to establish their safety and effectiveness before being licensed for use.
- The design of these clinical trials involves **randomised**, **double-blind** and **placebo-controlled protocols**, as shown in Figure 3.25.
- Subjects in clinical trials are divided into groups in a randomised way to reduce bias in the distribution of characteristics such as age and gender of the participants.
- In a double-blind trial, neither the subjects nor the researchers know which group subjects are in. This prevents bias. One group of subjects in the trial receives the vaccine or drug while the second group receives a placebo control to ensure valid comparisons.
- At the end of the trial, results from the two groups are compared to determine whether there are any statistically significant differences between the groups.
- Both groups must be of a suitable size to reduce the magnitude of experimental error.

Synoptic link

You can read more about vaccines in Key Area 3.7 (page 91).

Key terms

Clinical trial A method of obtaining data about new drugs, vaccines or other treatments.

Randomised trial A method of reducing bias in clinical trials by eliminating effects caused by variables such as age or gender of the participants.

Double-blind trial A clinical trial in which neither participants nor investigators know which participants are given the treatments and which are the control group.

Placebo-controlled protocol A type of control used in a clinical trial in which a random group of participants is given a 'blank' rather than the treatment under trial.

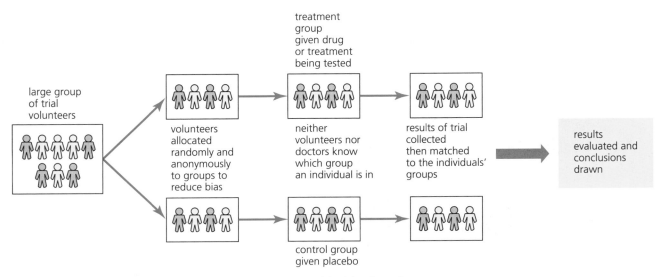

Figure 3.25 A randomised, placebo-controlled, double-blind trial

Do you know?

1 Write notes on how you would carry out a randomised, placebo-controlled, double-blind trial of a new vaccine.   [5]

Area 3 assessment
Neurobiology and immunology

Answer on separate sheets of paper. Mark your own work at: **hoddereducation.co.uk/ needtoknow/answers**

1 The nervous system can be divided into the central nervous system (CNS) and the peripheral nervous system (PNS).

   a Name the components of the CNS.   [1]

   b Name the division of the nervous system that is linked to the medulla.   [1]

   c Describe the difference between somatic and autonomic actions.   [2]

   d Give the term that describes the relationship between sympathetic and parasympathetic fibres.   [1]

2 Neurons connected to each other through synapses form neural pathways through the nervous system.

   a Describe a converging neural pathway.   [1]

   b Explain the benefits arising from a diverging neural pathway.   [2]

   c Name the neural pathway in which neurons later in the pathway link with earlier neurons, sending the impulse back through the pathway.   [1]

3 The brain is the complex central processing unit of the nervous system.

   a Give *one* function of the cerebral cortex.   [1]

   b State the function of the motor areas in the cerebral cortex.   [1]

   c State the function of the corpus callosum.   [1]

   d Give *one* example of an association area of the cerebral cortex.   [1]

4 Memories include past experiences, knowledge and thoughts.

  a Name *two* components of memory. [2]

  b Give *one* characteristic of short-term memory (STM). [1]

  c Explain how the process of 'chunking' can improve the capacity of STM. [1]

  d Explain the term 'working memory model'. [1]

  e Describe the serial position effect. [1]

  f Name *one* method for transferring information from STM to LTM. [1]

5 Neurotransmitters are the chemical messengers of the nervous system, and are used to transmit signals from one neuron to another.

  a Describe *one* function of glial cells. [1]

  b State the function of the myelin sheath. [1]

  c Give an account of the role of neurotransmitters at a synapse. [1]

  d Give *one* way in which the neurotransmitters are removed to prevent continuous stimulation of postsynaptic neurons. [1]

  e Name the neurotransmitter involved in reduction of pain intensity following a trauma. [1]

  f Describe the action of antagonistic drugs in the synaptic cleft. [1]

  g Describe how exposure to agonistic drugs can lead to drug tolerance. [1]

6 Non-specific body defences can be physical and chemical.

  a Describe the role of epithelial cells in the non-specific defence of the body. [1]

  b Describe the role of mast cells in the inflammatory response. [1]

  c Describe the role of phagocytes in overcoming bacterial infection. [2]

  d Describe the role of cytokines in cellular defence. [1]

7 Lymphocytes are white blood cells that are involved in the specific immune response.

  a Describe the role of B-lymphocytes in specific cellular defence of the body. [1]

  b Explain the cause of an allergic response. [1]

  c Describe the action of an antibody. [1]

  d Describe the action of T-lymphocytes in specific cellular defence against infection. [2]

  e State what happens in an autoimmune response. [1]

  f Describe *one* difference in the response of the immune system to a second exposure to the same pathogen. [1]

  g Explain why individuals with AIDS are more vulnerable to opportunistic infections. [1]

8 Immunisation describes the process whereby individuals are protected against infectious illness by the administration of a vaccine.

  a Describe what is meant by a 'vaccine'. [1]

  b Give *one* source of antigens that are used in vaccines to trigger an immune response. [1]

  c Explain the role of adjuvants in immunisation. [1]

  d Explain what is meant by the term 'herd immunity'. [1]

  e A different vaccine is required against each strain of the influenza virus. Explain why different vaccines are required. [1]

9 Medical research studies involving people are called clinical trials.

a Explain why new drugs and vaccines are subjected to clinical trials. [1]

b Clinical trials of vaccines use randomised, placebo-controlled protocols. Describe how these protocols are set up by the researchers. [2]

c Describe what is meant by a double-blind trial. [1]

d Explain why it is important to work with a large group of participants in a clinical trial. [1]

Extended response

10 Give an account of the nervous system under the following headings:

a the role of neurotransmitters at a synapse [5]

b the structure and function of neural pathways. [5]